ELECTRICIAN'S GUIDE TO THE BUILDING REGULATIONS

Covering Approved Document P:
Electrical safety – Dwellings and
including guidance on the Scottish
and Welsh Building Regulations

Published by The Institution of Engineering and Technology, London, United Kingdom

The Institution of Engineering and Technology is registered as a Charity in England & Wales (no. 211014) and Scotland (no. SC038698).

The Institution of Engineering and Technology is the institution formed by the joining together of the IEE (The Institution of Electrical Engineers) and the IIE (The Institution of Incorporated Engineers). The new Institution is the inheritor of the IEE brand and all its products and services, such as this one, which we hope you find useful.

First published 2005 (0 86341 463 X)
Reprinted (with minor amendments) 2006
Second edition 2008 (978-0-86341-862-4)
Reprinted (with minor amendments) 2009
Reprinted 2010, 2011, 2012
Third edition 2013 (978-1-84919-556-0)
Fourth edition (incl. Amendment No. 3 to BS 7671:2008) 2015 (978-1-84919-889-9)

Copies of this publication may be obtained from:
The Institution of Engineering and Technology
PO Box 96, Stevenage, SG1 2SD, UK
Tel: +44 (0)1438 767328
Email: sales@theiet.org
www.electrical.theiet.org/books

ISBN 978-1-84919-889-9 (wiro bound)

ISBN 978-1-84919-890-5 (electronic)

Typeset in the UK by the Institution of Engineering and Technology, Stevenage
Printed in the UK by A. McLay and Company Ltd, Cardiff

Contents

Cooperating organisations

The Institution of Engineering and Technology acknowledges the invaluable contribution made by the following organisations in the preparation of the Electrician's Guide to the Building Regulations.

Institution of Engineering and Technology

M. Coles BEng(Hons) MIET

P.E. Donnachie BSc CEng FIET

We would like to thank the following organisations for their continued support:

Association of Manufacturers of Domestic Electrical Appliances

BEAMA Installation

BEAMA Ltd

British Cables Association

Certsure trading as NICEIC and Elecsa

City & Guilds of London Institute

Department for Communities and Local Government

Electrical Contractors' Association

Electrical Contractors' Association of Scotland (SELECT)

Electrical Safety First

ERA Technology Ltd

EAL

The GAMBICA Association Ltd

Lighting Association

NAPIT

Scottish Government Building Standards Division

Society of Electrical and Mechanical Engineers serving Local Government

Electrician's Guide to the Building Regulations revised, compiled and edited by:

P. Cook CEng FIEE and T. Pickard IEng MIET

Preface

This book gives guidance on the Building Regulations for England, Scotland and Wales. It includes guidance not only on the requirements for electrical installations (Part P) but also for other parts of the Building Regulations (Parts A, B, C, E, F, L and M) that persons carrying out electrical installations are expected to comply with.

This book has been updated in line with BS 7671:2008+A3:2015 and the changes to Approved Document P that took place in April 2013. Further information pertaining to the third-party certification will be available from the IET website at **www.theiet.org/ building-regs** after the implementation of competency schemes by the Department for Communities and Local Government (DCLG). Check back to this web page to download the updates free of charge.

This book is supports a range of EAL qualifications, including:

▶ the EAL Level 3 Certificate In Installing, Testing and Ensuring Compliance of Electrical Installations in Dwellings (QCF); and
▶ the EAL Level 3 Award In Approving Electrical Installation Work in Dwellings in Compliance with Building Regulations

which have been developed to meet the new Qualified Supervisors technical competence requirements.

Domestic circuit schedules that comply with BS 7671:2008+A3:2015 *Requirements for Electrical Installations* (IET Wiring Regulations 17th Edition) are described. They are not necessarily the most cost-effective for any particular installation, but have the advantage of being simple and generally applicable.

Approved Documents are intended to provide guidance for some of the more common building situations. However, there may well be alternative ways of achieving compliance with the requirements. Thus, there is no obligation to adopt any particular solution contained in an Approved Document if you prefer to meet the relevant requirement in some other way.

Legislation 1

1.1 Health and safety

1.1.1 Health and Safety at Work etc. Act 1974

Apart from the common law general duty of care of everyone for their neighbour (including employees and anyone who might use the place of work), there is specific legislation with respect to safety at work, the most fundamental being the Health and Safety at Work etc. Act 1974.

The Health and Safety at Work etc. Act 1974 is comprehensive, and concerns health, safety and welfare at work, and the control of dangerous substances and certain emissions into the atmosphere. Sections 2, 3 and 4 of the Act place a duty of care upon the employer, the employee and the self-employed to ensure the health, safety and welfare at work of all persons, employees and others using the work premises.

This Act empowers the Secretary of State to make regulations. The most relevant to electrical installation work are:

▶ Electricity at Work Regulations 1989
▶ The Management of Health and Safety at Work Regulations 1999
▶ Construction (Design and Management) Regulations 2007
▶ Provision and Use of Work Equipment Regulations 1998
▶ Personal Protective Equipment at Work Regulations 2002
▶ Electricity Safety, Quality and Continuity Regulations 2002 (as amended).

NOTE: Lack of knowledge of a regulation is no defence in law.
(Further information on legislation relevant to persons engaged in electrical installation work is given in the IET publications *Commentary on the IET Wiring Regulations* and *Electrical Maintenance*.)

1.1.2 Electricity at Work Regulations 1989

The Electricity at Work Regulations (SI 1989 No 635) impose duties on every employer, every employee and every self-employed person to ensure that the safety requirements of the Regulations are met. The Regulations are enacted to provide for the electrical safety of persons in the workplace. The requirements of Regulation 4 are:

1   All systems shall at all times be of such construction as to prevent, so far as is reasonably practicable, danger.
2   As may be necessary to prevent danger, all systems shall be maintained so as to prevent, so far as is reasonably practicable, such danger.
3   Every work activity, including operation, use and maintenance of a system and work near a system, shall be carried out in such a manner as not to give rise, so far as is reasonably practicable, to danger.
4   Any equipment provided under these Regulations for the purpose of protecting persons at work on or near electrical equipment shall be suitable for the use for which it is provided, be maintained in a condition suitable for that use, and be properly used.

In practice, all electrical systems shall be designed, installed and maintained in use so as to prevent danger. The scope of the Electricity at Work Regulations includes not only the fixed wiring of the electrical installation but also current-using equipment supplied from the fixed installation, including appliances, computers, photocopiers, power drills, etc.

The *Memorandum of guidance on the Electricity at Work Regulations* published by the Health and Safety Executive (publication HSR25) provides guidance on all aspects of the Regulations. With respect to design and installation of electrical systems it says that '*The IEE Wiring Regulations* is a code of practice which is widely recognised and accepted in the UK and compliance with it is likely to achieve compliance with the relevant aspects of the 1989 Regulations'.

The scope of the Electricity at Work Regulations is much wider than BS 7671 *Requirements for Electrical Installations*, in that they require:

▶   installations to be constructed so as to be safe: Regulations 4(1), 5, 6, 7, 8, 9, 10, 11, 12;
▶   installations to be maintained so as to be safe: Regulation 4(2);
▶   associated work to be carried out safely: Regulations 4(3), 13, 14, 15;
▶   work equipment provided to be suitable for the purpose: Regulation 4(4);
▶   persons to be competent: Regulation 16.

The Electricity at Work Regulations are also concerned with good working practices.

Compliance with BS 7671 should provide for compliance with the Electricity at Work Regulations and the Building Regulations as they apply to the fixed installation. Compliance with the recommendations in the IET *Code of Practice for In-Service Inspection and Testing of Electrical Equipment* should provide for compliance with the Electricity at Work Regulations as they apply to current-using equipment such as appliances and electrical office equipment.

1.1.3 The Management of Health and Safety at Work Regulations 1999

The Management of Health and Safety at Work Regulations 1999 require every employer to manage health and safety aspects of their business including:

- ▶ carrying out risk assessments of the risks to the health and safety of employees whilst at work and to all persons affected by the work;
- ▶ appointing competent persons to assist in health and safety matters;
- ▶ establishing procedures to be followed in the event of serious danger; and
- ▶ arranging for appropriate information and training.

1.1.4 Construction (Design and Management) Regulations 2015

The Construction (Design and Management) Regulations 2015 apply generally to construction work, whether notifiable or not, including domestic installations. The basic requirement is that design and construction must take account of the health and safety aspects both in the construction phase of the work and during any subsequent maintenance. There are also requirements regarding the maintenance and repair of the construction at any time, including after construction work is completed, i.e. during use and during demolition.

There is a requirement to provide reasonably foreseeable information necessary for the health and safety of persons who will carry out maintenance, repairs and cleaning in the future. Consequently, persons with such responsibilities for a building should request access to the health and safety file that was prepared for the construction to see if there are any particular problems associated with the maintenance and repair of the building, including its electrical services.

Notification (F10) is required to the Health and Safety Executive (HSE) for projects lasting more than 500 person days, or lasting more than 30 working days with more than 20 workers, at any time, simultaneously on site.

Guidance is given in HSE publication L153: *Managing health and safety in construction – Construction (Design and Management) Regulations 2015 – Guidance on Regulations.* Domestic work is now within the scope of the Regulations and for domestic work the contractor assumes the duties of the client. (Downloadable at http://www.hse.gov.uk/pubns/books/l153.htm.)

Guidance specifically for clients is given in HSE publication INDG411: *Need building work done? A short guide for clients on the Construction (Design and Management) Regulations 2015.* (Downloadable at: http://www.hse.gov.uk/pubns/indg411.pdf.)

NOTE: Where the domestic project involves:

(a) **only one contractor**, the contractor must carry out the client duties as well as the duties they already have as a contractor.

(b) **more than one contractor**, the principal contractor must carry out the client duties as well as the duties they already have as principal contractor. If the domestic client has not appointed a principal contractor, the duties of the client must be carried out by the contractor in control of the construction work.

1.1.5 Provision and Use of Work Equipment Regulations 1998

The Provision and Use of Work Equipment Regulations 1998 require that work equipment, including installations, is so constructed (or adapted) as to be suitable for the purpose for which it is provided. Equipment must be inspected after installation and before use, at suitable intervals and after exceptional events, for example, fire, flooding, and mechanical damage. Potentially dangerous machinery must be guarded. Where necessary, logs must be maintained and adequate training given to operators.

1.1.6 Personal Protective Equipment at Work Regulations 2002

The Personal Protective Equipment at Work Regulations 2002 require every employer to ensure that suitable personal protective equipment is provided to employees, as may be necessary. The equipment must take account of:

▶ the risks;
▶ the environmental conditions;
▶ ergonomic requirements;
▶ the state of health of the person or persons; and
▶ comply with any appropriate provisions or standards.

There is a requirement upon employees to use protective equipment provided in accordance with training and instruction received. Employees are required to report the loss of such equipment.

The requirements of this legislation do mean that proper records should be kept of protective equipment and a suitable procedure set up for checking that employees still have the equipment necessary and that it is in good order. This does not in any way reduce the duty of the employee to advise the employer of any defects or deficiencies in the equipment or the training that he or she has received.

1.1.7 Electricity Safety, Quality and Continuity Regulations 2002 (as amended), Statutory Instrument No. 2665

The main purpose of the Electricity Safety, Quality and Continuity Regulations (ESQCR) is to provide for the safety of the electricity supply distribution system, i.e. for the safety of the general public and of persons working on the system. As the title implies, the Regulations also have requirements for power quality and supply continuity.

There are many references to BS 7671.

The Regulations require distributors prior to connection of new properties to obtain confirmation that the installation to be connected complies with the requirements of BS 7671 (regulations 9(4) and 25(2) of the ESQCR).

Regulation 28 requires the distributor to provide on request a written statement of:

(a) the maximum prospective short-circuit current at the supply terminals;

(b) for low voltage connections, the maximum earth loop impedance of the earth fault path outside the installation;

(c) the type and rating of the distributor's protective device or devices nearest to the supply terminals;

(d) the type of earthing system applicable to the connection; and

(e) the number of phases, frequency and voltage.

For new supplies, unless inappropriate for reasons of safety, regulation 24(4) requires the distributor to make available an earth connection. (This is almost always provided for in the UK by the provision of a PME (TN-C-S) supply.)

1.1.8 Construction (Health, Safety and Welfare) Regulations 1996

All the regulations have been revoked by the Construction (Design and Management) Regulations 2007.

1.2 The Building Regulations

Note: Approved documents and guidance can be freely downloaded from the DCLG website: http://www.planningportal.gov.uk

1.2.1 The Building Regulations 2010

England

The Building Regulations 2010 are made under the Building Act 1984 and apply to England. The aim of the Regulations is to ensure the health and safety of building users, promote energy efficiency, facilitate sustainable development and contribute to meeting the access needs of people in and around all types of buildings (that is, domestic, commercial and industrial).

Part P (Electrical safety – Dwellings) of the Building Regulations came into effect on 1 January 2005 with Statutory Instrument 2004 No. 3210, was amended on 5 April 2006 with Statutory Instrument 2006 No. 652 and amended again on 6 April 2013.

Wales

On 31 December 2011 the power to make building regulations for Wales was transferred to Welsh Ministers. This means that:

▶ Welsh Ministers will make any new building regulations or publish any new building regulations guidance applicable in Wales from that date; and

▶ the Building Regulations 2010 and related guidance, including approved documents as at that date, will continue to apply in Wales until Welsh Ministers make changes to them. For example the 2006 version of Approved Document P applies at the date of publication of the Guide.

As guidance is reviewed and changes made, Welsh Ministers will publish separate approved documents. The first of these will be changes to Part L (Conservation of fuel and power), for which consultations have taken place.

Please see Chapter 13 for guidance for contractors working in Wales.

Northern Ireland

In Northern Ireland the Building Regulations (Northern Ireland) 2000 (as amended) apply.

The Department of Finance and Personnel has produced Technical Booklet E:2005 and also refers to other deemed-to-satisfy publications, such as British Standards, to support compliance with the Building Regulations (Northern Ireland) 2000.

Scotland

The Building Regulations 2010 do not apply in Scotland.

In Scotland the requirements of the Building (Scotland) Regulations 2004 (as amended) apply, in particular regulation 9; see Chapter 12 for guidance.

In the Building Regulations 2010, 'building work' means:

▶ putting up a new building, or extending or altering an existing one (for example by converting loft space into living space);
▶ the insertion of insulation into a cavity wall;
▶ the underpinning of the foundations of a building;
▶ the erection or renovation of a thermal element (insulation), works associated with a change in energy status (no longer exempt from Part L), and consequential improvements; and
▶ providing services or fittings in a building, such as:

– washing and sanitary facilities (WCs, showers, washbasins, kitchen sinks, etc.);
– boilers and hot water cylinders;
– other combustion appliances of any type;
– oil and LPG fuel storage installations;
– foul water and rainwater drainage;
– replacement windows; and
– electrical work.

Certain changes of use to an existing building may result in the building as a whole no longer complying with the requirements that will apply to its new type of use.

Regulation 4 of the Building Regulations 2010 requires building work, as defined above, to comply with the performance requirements listed in Schedule 1, under the 14 Parts A to P. The clauses in regulation 4 most applicable to electrical installation work are:

4. **(1)** ..., building work shall be carried out so that:
  (a) it complies with the applicable requirements contained in Schedule 1; and
  (b) in complying with any such requirement there is no failure to comply with any other such requirement.

(2) Where –
  (a) building work is of a kind described in regulation 3(1)(g),(h) or (i); and
  (b) the carrying out of that work does not constitute a material alteration that work need only comply with the applicable requirements of Part L of Schedule 1.

(3) Building work shall be carried out so that, after it has been completed:
  (a) any building which is extended or to which a material alteration is made; or
  (b) any building in, or in connection with, which a controlled service or fitting is provided, extended or materially altered; or
  (c) any controlled service or fitting, complies with the applicable requirements of Schedule 1 or, where it did not comply with any such requirement, is no more unsatisfactory in relation to that requirement than before the work was carried out.

The primary responsibility for achieving compliance with the Building Regulations rests with the person carrying out the building work. If electrical installation work is non-compliant, the local authority will usually take action against the electrician. Alternatively, or in addition, the local authority may serve an enforcement notice on the owner of the building. Persons employing others to carry out work should confirm who has responsibility for compliance with the Regulations.

The Building Regulations define an electrical installation as 'fixed electric cables or fixed electrical equipment located on the consumer's side of the electricity supply meter'. Note that the definition in BS 7671 (*IET Wiring Regulations* 17th Edition) is different.

1.2.2 Approved documents

For the purposes of providing practical guidance with respect to the requirements of the different parts of the Building Regulations 2010 for England (and Wales), the Secretary of State has issued a series of approved documents.

This publication gives guidance on complying with Approved Document P: Electrical safety – Dwellings.

However, persons responsible for work within the scope of Part P of the Building Regulations may also be responsible for ensuring compliance with other parts of the Building Regulations where relevant, particularly if there are no other parties involved with the work. So advice is also provided in Chapter 10 on other Building Regulations requirements relevant to electricians carrying out electrical work, as follows:

Part A (Structure): Depth of chases in walls, sizes of holes and notches in floor and roof joists;

Part B (Fire safety): Fire safety of certain electrical installations; provision of fire alarm and fire detection systems; fire resistance of penetrations through floors and walls;

Part C (Site preparation and resistance to contaminants and moisture): Moisture resistance of cable penetrations through external walls;

Part E (Resistance to the passage of sound): Penetrations through floors, ceilings and walls;

Part F (Ventilation): Ventilation rates for dwellings;

Part G (Sanitation; hot water safety and water efficiency): Electric water heating

Part K (Protection from falling); Electrical means of opening windows

Part L (Conservation of fuel and power): Energy efficient lighting

Part M (Access to and use of buildings): Heights of switches, socket-outlets and consumer units

All the above are available from the Department for Communities and Local Government (DCLG) website: www.planningportal.gov.uk

The Building Regulations (regulation 4(3)) require that on completion of work in an existing building, the building should comply with the current Building Regulations, or, if it did not comply before the work began, be no worse in terms of the level of compliance with the other applicable Parts of Schedule 1 to the Building Regulations, including Parts A, B, C, E, F, L and M.

For example, holes cut in a ceiling for a recessed luminaire (light fitting) might degrade a floor's performance in terms of its resistance to fire (Part B) and sound penetration (Part E) and this is not allowed. Chasing of walls and drilling of joists must comply with Part A to ensure the building structure remains sound. Chapter 10 of this publication provides guidance.

1.2.3 Competent person self-certification schemes

The department for communities and local government require Competent person self-certification schemes to comply with their 'Conditions of authorisation'; see Appendix C. When these conditions are met the scheme is listed in Schedule 3A of the Building Regulations, see Appendix B, section 2.

1.3 Approved Document P: Electrical safety – Dwellings

1.3.1 The requirement

▼ **Figure 1.3.1** The requirement of Part P of the Building Regulations

Requirements	Limits on application
Design and installation	
P1 Reasonable provision shall be made in the design and installation of electrical installations in order to protect persons operating, maintaining or altering the installations from fire or injury.	The requirements of this part apply only to electrical installations that are intended to operate at low or extra-low voltage and are – **(a)** in or attached to a dwelling; **(b)** in the common parts of a building serving one or more dwellings, but excluding power supplies to lifts; **(c)** in a building that receives its electricity from a source located within or shared with a dwelling; or **(d)** in a garden or in or on land associated with a building where the electricity is from a source located within or shared with a dwelling.

1.3.2 Scope of Part P

Part P applies to electrical installations:

   (a) in a dwellinghouse or flat, and to parts of the installation that are:

      (i) outside the dwelling – for example, fixed lighting and air conditioning units attached to outside walls, photovoltaic panels on roofs, and fixed lighting and pond pumps in gardens;

      (ii) in outbuildings, such as sheds, detached garages and domestic greenhouses; and

   (b) in the common access areas of blocks of flats such as corridors and staircases;

   (c) in shared amenities of blocks of flats such as laundries, kitchens and gymnasiums; and

   (d) in business premises (other than agricultural buildings) connected to the same meter as the electrical installation in a dwelling – for example, shops and public houses below flats.

Part P does not apply to electrical installations:

   (a) in business premises in the same building as a dwelling with separate metering; and

(b) that supply the power for lifts in blocks of flats (but Part P does apply to lift installations in single dwellings).

NOTE: Schedule 2 to the Building Regulations identifies buildings – for example, unoccupied, agricultural, temporary and small detached buildings – that are generally exempt from the requirements of the Regulations. However, conservatories, porches, domestic greenhouses, garages and sheds that share their electricity with a dwelling are not exempt from Part P (by virtue of regulation 9(3)) and must comply with its requirements.

See Figure 1.3.2.

▼ **Figure 1.3.2** Scope of Part P

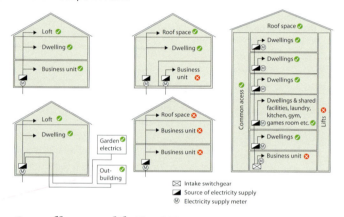

1.3.3 Compliance with Part P

In the Secretary of State's view, the requirements of Part P will be met if low voltage and extra-low voltage electrical installations in dwellings are designed and installed so that both of the following conditions are satisfied:

(a) they afford appropriate protection against mechanical and thermal damage; and

(b) they do not present electric shock and fire hazards to people.

Electrical installations should be designed and installed in accordance with BS 7671:2008 incorporating Amendment No. 3:2015.

This publication is written to provide simple rules and installation requirements including circuit specifications for compliance with BS 7671 and other relevant parts of the Building Regulations.

1.3.4 Parts of the Building Regulations other than Part P

The requirements of other applicable parts of the regulations may affect the electrical installation work. The 'other' relevant part must be complied with (see Chapter 10) or be no less compliant on completion of the work than before the work was started.

1.4   Notification to Building Control, etc.

1.4.1   Prior notification

Except as below, the relevant building control body must be notified of all proposals to carry out electrical installation work in dwellings, etc. (see section 1.3) before the work begins.

It is not necessary to give prior notification of proposals to carry out electrical installation work in dwellings to building control bodies if the work:

(a) is carried out and certified by a registered competent enterprise; or

(b) is certified by a registered third party certifier; or

(c) is non-notifiable work.

The technical requirements of Part P apply to all electrical installation work in dwellings, whether the work needs to be notified to Building Control or not.

1.4.2   Registered competent person

Electrical installers who are registered electrically skilled persons should complete a BS 7671 type electrical installation certificate for every job they undertake. The electrical installer should give the certificate to the person ordering the work.

The installer or the installer's registration body must do both of the following within 30 days of the work being completed:

(a) Give a copy of the Building Regulations compliance certificate to the occupier.
   NOTE: In the case of rented properties, the certificate may be sent to the person ordering the work and copied to the occupier.

(b) Give the certificate, or send a copy of the information on the certificate, to the building control body (see Figure 1.4.2).

1.4.3   Registered third party certifier

Before the work begins, an installer who is not a registered electrically skilled person may appoint a registered third party certifier to inspect and test the work as necessary.

Within 7 days of completing the work, the installer must notify the registered third party certifier who, subject to the results of the inspection and testing being satisfactory, should then complete an electrical installation condition report and give it to the person ordering the work.

The third party electrical installation condition report should be a model form, or one developed specifically for Part P purposes. If the third party certifier is satisfied that the work complies with the Building Regulations, the third party certifier or the registration body of the registered third party certifier must do both of the following within 30 days:

(a) give a copy of the Building Regulations compliance certificate to the occupier; and

(b) give a certificate, or a copy of the information on the certificate, to the building control body.

If the third party certifier is unable to certify that the requirements of the Building Regulations have been met, the third party certifier must notify the local authority.

▼ **Figure 1.4.2** Building Regulations notification

NAPIT Building Regulations Part P
Postal Notification

Domestic /associated non dwelling work notification to NAPIT

NAPIT Company Membership No. | 1 4 7 8 9 0 1 — Electrical Certificate Number | 6 7 2 5 6 2 4 6 8 9 2 0

1 Name of NAPIT member/Company who carried out the electrical installation

SAMPLE

Address for Notification certificate to be sent, if different from the address where the work has been carried out

MR J SMITH
SMITH AND SONS LTD
24 STATION ROAD
DERHAM
NORFOLK
Postcode *NR19 7XE*

IT'S ESSENTIAL TO GIVE US THIS

Installation Address
Address
48 HIGH STREET, OLD TOWN
County *HERTFORDSHIRE*
Date the work was completed *XX/XX/XX*
Client name *MR J SMITH*
New build address of plot number [eg plot17] *PLOT17*
Local authority [if known]

Postcode *SG15 9AE*

(IT IS ESSENTIAL TO GIVE US THE POSTCODE OTHERWISE THE FORM WILL BE RETURNED)

NOTIFIABLE WORK DETAILS

2 Please give number of circuits / modifications done in each area. →

	House Dwelling	Flat Dwelling	Common Area of Block of flats	Building Sharing Supply with Dwelling	Building extension or conservatory	Detached Shed Garage or Greenhouse	Kitchen	Special Location, Bath, Shower, Pool, Sauna	Garden
Install one or more new circuits									
Install a replacement consumer unit									
Rewire of all circuits									
Partial rewire									
New full electrical installation									
New full electrical installation (new build)									
Circuit alteration or addition in a special location									
Install an electric storage heater in a dwelling (PAS 2030 measured only)									
Install an electric storage heater in a non-dwelling (PAS 2030 measured only)									
Install an electric storage heater, with warm air heat distribution in dwelling (PAS 2030 measured only)									
Install an electric storage heater, with warm air heat distribution in non-dwelling (PAS 2030 measured only)									
Installation of non domestic energy efficient electrical heating system									
Installation of light fitting, systems and, system controls in a dwelling (PAS 2030 measured only)									
Installation of light fitting, systems and, system controls in a non-dwelling (PAS 2030 measured only)									
Installation of non domestic energy efficient lighting									
Install a variable speed drive for fans and pumps in a non-dwelling (PAS 2030 measured only)									
Install a low energy circulator pump (PAS 2030 measured only)									

DESCRIPTION OF LOCATION

3 Declaration The above electrical installation complies with building regulations part 4 and is designed, installed, inspected, tested and certified in accordance *2015* year

Signature *P. JAMESON* Date *X X XXXX* Must be no more than 14 days after completion of work

Print name *PETER JAMESON*

Send to NAPIT Administration Centre, 4th Floor, Mill 3, Pleasley Vale Business Park, Mansfield, Nottinghamshire NG19 8RL

Sheet 1 of 1 NA/01010A (V5)

1.4.4 Conditions of authorisation of third party certification schemes

The Conditions of authorisation of third party certification schemes have been loaded on to the UK government planning portal.

1.4.5 Certification by a building control body

If the work is not to be carried out by a registered electrically skilled person, or supervised, including inspection and testing, by a registered third party certifier, then before starting work the person responsible for the work must notify the building control body.

The building control body will determine the extent of inspection and testing needed for it to establish that the work is safe, based on the nature of the electrical work and the competence of the installer. The building control body may choose to inspect and test the electrical work itself, or to contract out some or all of the work to a specialist.

An installer who is competent to carry out inspection and testing should give the appropriate BS 7671 certificate to the building control body, who will then take this certificate and the installer's qualifications into account when deciding what further action (if any) it needs to take. Building control bodies may ask installers for evidence of their qualifications. Under the Building (Local Authority Charges) Regulations 2010, a local authority is required to take account of the receipt of a BS 7671 certificate from a person competent to carry out inspection and testing in setting its building control charge.

1.4.6 Unregistered persons

Persons not registered with an assessed enterprise are required to notify Building Control of notifiable work before work starts (emergency work is to be notified as soon as possible). Inspection and testing should be carried out as per BS 7671. Completed certificates with schedules should be forwarded to Building Control who will take the certificates into account when deciding what further action, if any, needs to be taken.

Building Control on notification become responsible for the work. If Building Control decide the completed work is safe and meets the requirements of all the Building Regulations, they will issue a Building Regulations completion certificate on request.

Members of bodies such as the Institution of Engineering and Technology (IET) who carry out electrical work in their own homes are not exempt from the requirement to notify Building Control, in the same way that members of the Institution of Civil Engineers who do work on their house foundations are not exempt.

1.4.7 Unqualified installers

Installers (contractors or DIYers) not qualified to inspect and test their work must notify the building control body of notifiable work before the work starts (emergency work is notified as soon as possible). Building control is then responsible for ensuring the work is safe, which includes arranging for inspection and testing as necessary.

1.4.8 Compliance certificates

On completion (including inspection and testing) of notifiable work, the occupier is to receive:

- ▶ a Building Regulations compliance certificate (issued by the self-certification scheme on behalf of the registered competent enterprise); or
- ▶ a building control completion certificate issued by the building control of the local authority; or
- ▶ a final report issued by third party certifier approved inspectors; and
- ▶ an appropriate electrical installation certificate complete with schedules.

1.4.9 Non-notifiable work

Non-notifiable electrical installation work (see section 1.6) should be carried out in accordance with BS 7671. If local authorities find that non-notifiable work is unsafe and non-compliant, they can take enforcement action.

1.5 Notifiable work

1.5.1 Requirement P1

Any electrical work carried out in a dwelling is subject to Requirement P1 and should be designed and installed to comply with BS 7671:2008 including Amendment No. 3: 2015.

1.5.2 Notifiable work

Regulation 12(6A) of the Building Regulations 2010 describes notifiable work as follows:

Notifiable work

A person intending to carry out building work in relation to which Part P of Schedule 1 imposes a requirement is required to give a building notice or to deposit full plans where the work consists of –

(a) the installation of a new circuit;
(b) the replacement of a consumer unit; or
(c) any addition or alteration to existing circuits in a special location

'special location'* means –

(a) within a room containing a bath or shower, the space surrounding a bath tap or shower head, where the space extends –
   (i) vertically from the finished floor level to –
      (aa) a height of 2.25 metres; or
      (bb) the position of the shower head where it is attached to a wall or ceiling at a point higher than 2.25 metres from that level; and
   (ii) horizontally –
      (aa) where there is a bath tub or shower tray, from the edge of the bath tub or shower tray to a distance of 0.6 metres; or

> **(bb)** *where there is no bath tub or shower tray, from the centre point of the* shower head where it is attached to the wall or ceiling to a distance of 1.2 metres; or
>
> **(b)** a room containing a swimming pool or sauna heater.

***NOTE:** The term 'special location' used in Approved Document P is not the same as that used in BS 7671, it is limited as stated in 1.5.2 to a 'room containing a sauna heater, or a swimming pool and the space surrounding a bath or shower'.

▼ **Figure 1.5.2** Extent of the 'special location' in a room containing a bath or shower

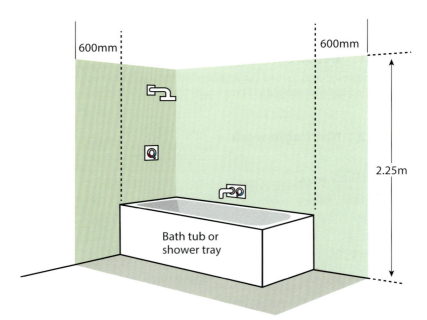

600mm   600mm

2.25m

Bath tub or shower tray

Additions and alterations to existing circuits are notifiable in the shaded area

NOTE: Socket-outlets should not be located within 3m of a bath tub or shower tray

1.6   Non-notifiable work

Whilst all fixed electrical installation work associated with a dwellinghouse or flat is within the scope of Part P and must comply with the *IET Wiring Regulations* (BS 7671), only work described in paragraph 1.5.2 is notifiable, the rest is not.

Examples of non-notifiable work:

(a) adding socket-outlets to an existing circuit other than in a Building Regulations 'special location';

(b) adding lighting points to an existing circuit other than in a Building Regulations 'special location';

(c) installing an outside light or socket supplied from an existing circuit;

(d) installing lights and sockets in a conservatory from existing circuits; and

(e) installing socket-outlets or lighting points from existing circuits in a kitchen.

It must be noted that non-notifiable work is required to comply with BS 7671 including the requirements for inspection and testing and the issuing of a minor works certificate.

BS 7671 certificates can usually only be issued by the person who carried out the electrical work. The guidance in Approved Document P, however, does allow DIYers undertaking non-notifiable work to have their work checked by a competent third party.

1.7   Provision of information

After carrying out work, including a new installation, sufficient information shall be provided to the person ordering the work for passing on to the occupant so that persons wishing to operate, maintain or alter an electrical installation can do so with reasonable safety.

Meeting the requirements of BS 7671 will require the installer to:

(a) provide a schedule of inspections, schedule of tests, and electrical installation certificate or periodic inspection certificate (or minor works certificate if appropriate) – see Chapter 6;

(b) provide labelling of the installation as in section 3.4;

(c) install cables in the building fabric only as permitted in section 2.3; and

(d) provide operating instructions and logbooks, and for large or complex installations, detailed plans.

1.8   Scottish Building Regulations

Chapter 12 explains the requirements for electrical installations of the Building (Scotland) Act 2003 and associated legislation.

1.9   Welsh Building Regulations

Welsh Ministers are now responsible for remaining functions under the Building Act 1984 – including making Building Regulations.

The functions transferred are limited under the Welsh Ministers (Transfer of Functions) (No. 2) Order 2009 (S.I. 2009/3019).

Documents and guidance that were current and in force up to 31 December 2011 and which had previously applied to England and Wales will for the time being continue to apply in Wales following transfer of powers. This includes:

    (a) Building Regulations;
    (b) the Approved Documents;
    (c) guidance published by the Department for Communities and Local Government (DCLG); and
    (d) approvals under competent persons provisions.

As of 1 January 2012, any revisions to Building Regulations and related procedures, processes and guidance proposed or issued by DCLG apply to England only.

See Chapter 13 for further guidance.

1.10 Building Regulations in Northern Ireland

http://www.buildingcontrol-ni.com/

The new Building Regulations were made on 15 May 2012 and came into operation on 31 October 2012.

The Northern Ireland Building Regulations are legal requirements made by the Department of Finance and Personnel and administered by 26 District Councils. The Regulations are intended to ensure the safety, health, welfare and convenience of people in and around buildings. They are also designed to further the conservation of fuel and energy.

Guidance on how to do this can be found in Technical Booklets prepared by the Department of Finance and Personnel. Adherence to the methods and standards detailed in the Technical Booklets means that the work will be 'deemed-to-satisfy' and must be accepted by Building Control as complying with the relevant Regulations.

A designer or builder can, however, use other methods, provided it can be demonstrated that the requirements of the Regulations have been met.

Design, selection and erection of electrical installations 2

NOTE: The calculation of load current from kVA and kW is described in sections 9.6 to 9.9.

2.1 Design

2.1.1 New buildings (installations)

The Electricity Safety, Quality and Continuity Regulations 2002 (as amended) (ESQCR) require the electricity distributor to install the cut-out and meter in a safe location, where they are mechanically protected and can be safely maintained. In compliance with this requirement, the electricity distributor and installer may be required to take into account the risk of flooding (some guidance is given in the DCLG publication *Preparing for Floods*, available from www.planningportal.gov.uk).

In accordance with the ESQCR 2002 (as amended) and the contract for a mains supply, proposals for new installations or significant alterations to existing ones must be agreed with the electricity distributor.

All new electrical installations including those in dwellings must comply with BS 7671. This guide provides uncomplicated examples of installations that will comply with BS 7671.

2.1.2 Additions and alterations to an installation

2.1.2.1 Electrical

BS 7671 Regulation 132.16 requires that no addition or alteration, temporary or permanent, shall be made to an existing installation, unless it has been ascertained that the rating and the condition of any existing equipment, including that of the distributor, will be adequate for the altered circumstances and that the earthing and bonding arrangements are also adequate.

When carrying out an addition or alteration to an existing electrical installation, the installer must confirm that the old installation meets the current requirements of the Building Regulations in so far as it affects the new installation and is adequate in all respects to supply the additional load, if any, such that the new work is safe, i.e. complies with Part 1 of BS 7671 *Requirements for Electrical Installations*.

In particular, the rating and the condition of the existing equipment belonging to both the customer and the electricity distributor can:

▶ carry the additional load;
▶ provide adequate shock protection (fault loop impedances must be appropriate for the protective devices); and
▶ the earthing and equipotential bonding arrangements are satisfactory.

Defects in the existing installation affecting the addition or alteration must be made good. Defects in the electrical installation identified during the work and not affecting the addition or alteration must be identified in writing to the person ordering the work.

2.1.2.2   Building works

Additions or extensions to a building must comply with the Building Regulations. If wiring is carried out in an extension to an existing building, the wiring to the extension must comply not only with Part P but also with all the other appropriate requirements of the Building Regulations. There is no particular requirement to ensure compliance with the Building Regulations with respect to other parts of the building not affected by the extension, alteration or addition. The compliance with the Building Regulations of the existing building must not be degraded. For example, energy-efficient luminaires must not be replaced by less efficient luminaires – however insistent the customer may be. Walls and ceilings must not be degraded by chases, drillings, etc. (see section 10.1 of this Guide for details).

2.1.2.3   Accessibility

Wall-mounted switches, socket-outlets and consumer units should be located so that they are easily reachable where this is necessary to comply with Part M (Access to and use of buildings) of the Building Regulations (see section 10.7). Consumer units should not be installed where young children might interfere with them. See section 3.1.

2.2 Selection of materials

2.2.1 The Building Regulations

Regulation 7 of the Building Regulations requires that building work shall be carried out:

(a) with adequate and proper materials that:
    (i) are appropriate for the circumstances in which they are used;
    (ii) are adequately mixed or prepared; and
    (iii) are applied, used or fixed so as to adequately perform the functions for which they are designed; and

(b) in a workmanlike manner.

2.2.2 Compliance with equipment standards

Part 1 of BS 7671 requires that every item of equipment must comply with the appropriate Harmonized Standard (EN) or Harmonized Document (HD) or National Standard implementing the HD. In the absence of an EN or HD the equipment must comply with the appropriate National Standard. In the UK an EN is published as a 'BS EN' and, where a Harmonization Document (HD) exists, it will be incorporated into the National Standard (BS). In selecting equipment, care must be taken to ensure that there is a declaration from the supplier that the equipment complies with the appropriate BS.

When selecting equipment, check that it is marked with the relevant BS number from the list in Appendix 1 of BS 7671.

It is a requirement of the Electrical Equipment Safety Regulations that before electrical equipment is put on sale, the manufacturer or authorised representative confirms that it complies with the safety requirements of the regulations. The manufacturer or authorised representative confirms that the equipment complies with all the requirements of the appropriate European directives and indicates this by applying the CE mark.

▼ **Figure 2.2.2**   CE mark

2.2.3 Independent certification schemes

For third party assurance that equipment complies with the appropriate BS, purchasers can look for an approval body mark such as the BSI Kitemark or the ASTA/BEAB mark and for cables the British Approvals Service for Cables (BASEC) mark.

BSI Kitemark
BSI has tested the product and has confirmed that the product conforms to the relevant British Standard

HAR mark
European third-party certification mark for cables and cords complying with relevant European safety standards (ENs/HDs)

British Approvals Service for Cables mark

ENEC mark
European third-party certification mark for electrical equipment complying with the European safety standards, for luminaires, transformers, power supply units and switches

ASTA Diamond mark
Tested and conforms to standard and factory quality management to ISO 9001

CEN/CENELEC Keymark
European third-party certification mark for household and similar electrical appliances, complying with relevant European safety standards

BEAB Approved mark
Electrical safety mark for household and similar appliances

2.3   Installation

2.3.1   Cable installation methods

The maximum conductor operating temperature that a cable can withstand limits the current rating of the cable. The conductor temperature is determined by:

(a) the current; and

(b) the thermal conductivity of the cable and its surroundings, particularly thermal insulation.

As a result, the installation method of a cable affects its current-carrying capability. Standard installation reference methods have been determined and when selecting a standard circuit from Chapter 4 not only must the type of fuse or circuit-breaker be known, but also the installation reference method. For circuit cables with more than one installation reference method, that with the lowest current rating is the reference method used in the calculation. The reference methods considered, in current rating order, are:

- ▶ (C) clipped direct to the surface, or embedded in plaster, see Figure 2.3.1a
- ▶ (B) installed in conduit or trunking on the surface, see Figures 2.3.1b, 2.3.1c and 2.3.1d
- ▶ (B) installed in conduit embedded in plaster, masonry or the like, see Figure 2.3.1e
- ▶ (100) above a plasterboard ceiling covered by thermal insulation not exceeding 100 mm, see Figure 2.3.1f
- ▶ (102) in a stud wall with thermal insulation with the cable touching the wall surface, see Figure 2.3.1h
- ▶ (A) installed in conduit or trunking in a thermally insulated wall
- ▶ (101) above a plasterboard ceiling covered by thermal insulation exceeding 100 mm, see Figure 2.3.1g
- ▶ (103) totally enclosed in thermal insulation for more than 0.5 m, see Figure 2.3.1i.

Method C has the highest current rating and 103 the lowest, see Table 2.3.1.

NOTE: Cables should not be totally surrounded by thermal insulation. If cables are totally surrounded by thermal insulation, cable sizes may need to be increased and this may result in practical problems when terminating the cables in accessories. For this reason it is recommended that, where practicable, precautions are taken to prevent cables being totally enclosed.

The installation reference methods are described below.

Reference method C

Sheathed cables, armoured or unarmoured, clipped direct or embedded in plaster.

▼ **Figure 2.3.1a**

Reference method B

Single-core or insulated and sheathed cables run in conduit or trunking installed on or in plaster, masonry or the like.

▼ **Figure 2.3.1b**

▼ **Figure 2.3.1c**

▼ **Figure 2.3.1d**

▼ **Figure 2.3.1e**

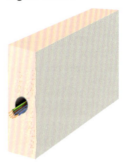

Reference method 100

Thermoplastic (PVC) insulated and sheathed flat cables above a plasterboard ceiling where the cables are touching the ceiling or clipped to joist covered by thermal insulation not exceeding 100 mm.

▼ **Figure 2.3.1f** Reference method 100

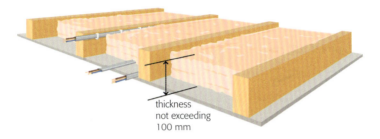

thickness
not exceeding
100 mm

Reference method 101

Thermoplastic (PVC) insulated and sheathed flat cables above a plasterboard ceiling where the cables are touching the ceiling or clipped to joist covered by thermal insulation exceeding 100 mm.

▼ **Figure 2.3.1g** Reference method 101

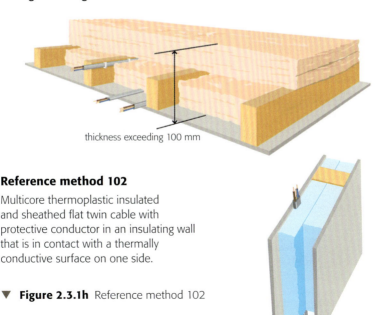

thickness exceeding 100 mm

Reference method 102

Multicore thermoplastic insulated and sheathed flat twin cable with protective conductor in an insulating wall that is in contact with a thermally conductive surface on one side.

▼ **Figure 2.3.1h** Reference method 102

Reference method 103

If cables are surrounded by thermal insulation for less than 5 cm, no de-rating is necessary. Where cables are totally enclosed over a length of 0.5 m or more the rating is half that for cables clipped directly to a conducting surface and unenclosed (reference method C). See Regulation 523.9 and Table 52.2 of BS 7671.

It is preferable for the installation of cables to be so arranged that the cables are not totally enclosed.

▼ **Figure 2.3.1i** Reference method 103

▼ **Table 2.3.1** Cable ratings (in amperes) for 70 °C thermoplastic insulated and sheathed flat cables with protective conductor (from Table 4D5 of BS 7671, except B)

	Installation reference method	Conductor cross-sectional area (mm²)						
		1.0	1.5	2.5	4	6	10	16
C	Clipped direct	16	20	27	37	47	64	85
B	Enclosed in conduit or trunking on a wall, etc.	13	16.5	23	30	38	52	69
	(from Table 4D2A of BS 7671)							
100	In contact with plasterboard ceiling or joists covered by thermal insulation not exceeding 100 mm	13	16	21	27	34	45	57
102	In a stud wall with thermal insulation with cable touching the wall	13	16	21	27	35	47	63
A	Enclosed in conduit in an insulated wall	11.5	14.5	20	26	32	44	57
101	In contact with plasterboard ceiling or joists ceiling covered by thermal insulation exceeding 100 mm	10.5	13	17	22	27	36	46
103	Surrounded by thermal insulation including in a stud wall with thermal insulation with cable not touching a wall	8	10	13.5	17.5	23.5	32	42.5

2.3.2 Floors and ceilings

When a cable is installed under a floor or above a ceiling it must be run in such a position that it is not liable to damage by contact with the floor or ceiling or their fixings. Unarmoured cables passing through a joist shall be at least 50 mm measured vertically from the top or bottom as appropriate or enclosed in an earthed steel conduit. Alternatively, the cables can be provided with mechanical protection sufficient to prevent penetration of the cable by nails, screws and the like. (**Note:** the requirement to prevent penetration can be difficult to meet.) (See also section 10.1 regarding Approved Document A.)

▼ **Figure 2.3.2**   Cables through joists

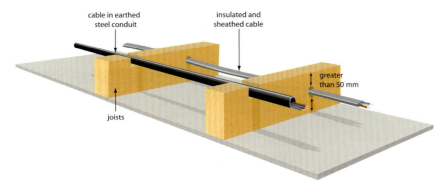

NOTES:

(a)   Maximum diameter of hole should be 0.25 × joist depth.
(b)   Holes on centre line in a zone between 0.25 and 0.4 × span.
(c)   Maximum depth of notch should be 0.125 × joist depth.
(d)   Notches on top in a zone between 0.07 and 0.25 × span.
(e)   Holes in the same joist should be at least 3 diameters apart.

2.3.3   Walls

A cable installed in a wall or partition must be run in such a position or otherwise protected that it is not liable to be damaged by contact with the wall structure or construction fixings. This is particularly important when the wall or partition is metal or part metal construction. Care must be taken to ensure that when wall cladding, door frames, etc. are fixed, cables are not damaged and metal structures made live.

NOTE: A zone formed on one side of a partition wall of 100 mm or less thickness extends to the reverse side only if the location of the accessory can be determined from the reverse side.

Where a cable is concealed in a wall or partition at a depth of less than 50 mm from any surface it must:

(a) have protection against penetration of the cable by enclosure in earthed metal conduit (trunking or ducting), or similar; or

(b) be installed either horizontally within 150 mm of the top of the wall or partition or vertically within 150 mm of the angle formed by two walls; or

(c) run horizontally or vertically to an accessory or consumer unit (see Figure 2.3.3); or

(d) form part of a SELV or PELV circuit.

Where indents (b) or (c) apply, but not (a) or (d), the cable shall be provided with additional protection by means of a 30 mA RCD.

▼ **Figure 2.3.3** Permitted cable routes

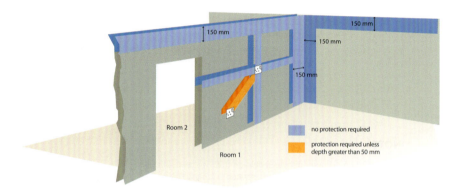

NOTES:

Except for SELV and PELV circuits, 30 mA RCD protection is required when a cable is:

(a) installed in a wall or partition the construction of which includes metal parts (fixings such as nails or screws are not included) and not protected by steel conduit or trunking; or

(b) installed in a wall or partition at a depth of less than 50 mm and not enclosed in steel conduit.

Mains position 3

3.1 Location and accessibility

3.1.1 The Electricity Safety, Quality and Continuity Regulations

ESQCR requires the electricity distributor to install the cut-out and meter in a safe location, where they are mechanically protected and can be safely maintained.

In compliance with this requirement, the electricity distributor and installer may be required to take into account the risk of flooding. Distributor's equipment and the installation consumer unit/fuseboard should be installed above the flood level. Upstairs power and lighting circuits and downstairs lighting circuits should be able to be installed above the flood level. Upstairs and downstairs circuits should have separate overcurrent devices (fuses or circuit-breakers).

Consumer units should not be installed where young children might interfere with them.

3.1.2 Consumer units

Enclosure material

Consumer units and other switchgear assemblies are required to be made of non-combustible material i.e. ferrous metal (steel). As an alternative, the consumer unit may be enclosed in a cabinet constructed from non-combustible material. See section 2.2.6 of *On-Site Guide*.

Meter tails

All the meter tails must enter a ferrous enclosure through the same entry point. They must be securely fixed/protected to prevent them being pulled out of their terminations.

Accessibility

Consumer units should not be installed where young children might interfere with them.

Approved Document P advises:

'Approved Document M does not recommend a height for new consumer units. However, one way of complying with Part M in new dwellings is to mount consumer units so that the switches are between 1350 mm and 1450 mm above floor level. At

this height, the consumer unit is out of reach of young children yet accessible to other people when standing or sitting.'

3.1.3 Separation of gas installation pipework from other services

Where installation pipework is not separated from electrical equipment or cables by an insulating enclosure, dividing barrier, trunking or conduit, it shall be spaced as follows:

(a) at least 150 mm away from electricity meters, controls, electrical switches or sockets, distribution boards or consumer units;

(b) at least 25 mm away from electricity cables.

The installation pipework shall not be positioned in a manner that prevents the operation of any electrical accessory, i.e. a switch or socket outlet.

NOTE: Where these spacing requirements are impracticable the pipework should either be sheathed with an electrical insulating material rated at 230 V a.c. or more, or a panel of electrical insulating material should be interposed.

(BS 6891:2014 *Specification for the installation and maintenance of low pressure gas installation pipework of up to 35mm (R11/4) on promises,* clause 8.4.2.)

▼ **Figure 3.1.1**   Separation from gas service pipes

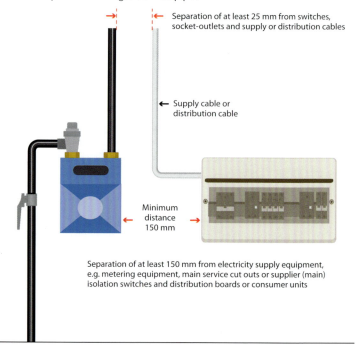

Separation of at least 25 mm from switches, socket-outlets and supply or distribution cables

Supply cable or distribution cable

Minimum distance 150 mm

Separation of at least 150 mm from electricity supply equipment, e.g. metering equipment, main service cut outs or supplier (main) isolation switches and distribution boards or consumer units

3.2   Supply systems

ESQCR requires the electricity distributor (regulation 27) to advise of:

(a) the number of phases;
(b) the frequency; and
(c) the voltage;

and, on request, the distributor (regulation 28) to provide the following information:

(d) the maximum prospective short-circuit current at the supply terminals;
(e) for low voltage connections, the maximum earth loop impedance of the earth fault path outside the installation;
(f) the type and rating of the distributor's protective device or devices nearest to the supply terminals;
(g) the type of earthing system applicable to the connection; and
(h) requirements for cross-sectional area and maximum length of meter tails.

This Guide assumes:

(a) single-phase supply; with
(b) 50 cycles per second frequency; and
(c) 230 V nominal voltage to earth;
(d) the maximum prospective short-circuit current is 16 kA at the supply terminals;
(e) the maximum external earth fault loop impedance (Z_e) is 0.35 ohm for PME (TN-C-S) supplies and 0.8 ohm for cable sheath earth supplies (TN-S); and
(f) a distributor's maximum protective device (cut-out fuse) rating of 100 A.

See below for earthing systems.

3.2.1   Protective multiple earthing (PME) supplies (TN-C-S system)

Almost all new supplies to dwellings will be from PME distribution systems (see Figure 3.2.1). The feature of such supplies is that the means of earthing for the installation is provided from the distributor's fused cut-out, where it is common with the PEN or neutral conductor. The size of the earthing and bonding conductors for standard domestic dwellings is given on the drawing. For non-standard installations, particularly larger installations, the sizes are tabulated (see Table 3.3.2a).

Except in city centres the conditions assumed are that for a TN-C-S system:

▶ the maximum external earth fault loop impedance, Z_e, is 0.35 ohm; and
▶ the maximum prospective fault current is 16 kA.

▼ **Figure 3.2.1** PME supply (TN-C-S system). Schematic of earthing and main protective bonding arrangements. Based on 25 mm² tails and selection from Table 54.7 of BS 7671.

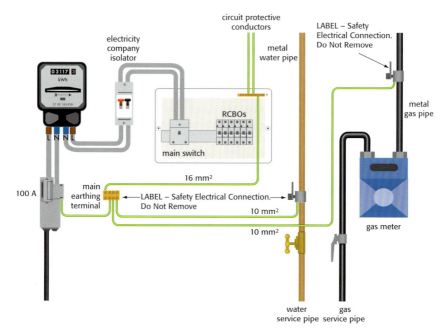

NOTES:

(a) An isolator is not always installed by the electricity distributor.

(b) Distributors will quote a Z_e of 0.35 Ω and a fault level of 16 kA. A loop impedance of 0.35 Ω equates to a fault level of 230 V/0.35 Ω = 657 A. 16 kA is the worst-case short-circuit fault level and 0.35 Ω is the worst-case external earth loop impedance.

3.2.2 Cable sheath earth (TN-S system)

Cable sheath earths are identifiable by the earth connection being made to the cable sheath. This earth connection should be connected on to the sheath.

The distributor is responsible for this connection, which should be securely and reliably made by a means such as soldering or brazing.

A maximum fault level of 16 kA may be assumed and a maximum external earth loop impedance of 0.8 ohm.

▼ **Figure 3.2.2** Cable sheath earth (TN-S system). Schematic of earthing and main protective bonding arrangements. Based on 25 mm² tails and selection from Table 54.7 of BS 7671.

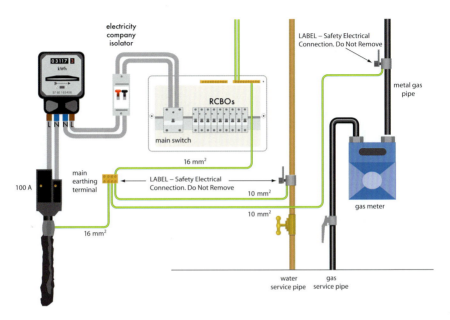

NOTE: An isolator is not always installed by the electricity distributor.

3.2.3 No earth provided (TT system)

TT installations will often be encountered in rural areas, where there are overhead supplies. Additionally, an installation forming part of a TT system may be encountered where a distributor may not be prepared to provide an earthing terminal for an installation, such as that for a swimming pool, farm or building site. These installations are generally outside the scope of this Guide.

Figure 3.2.3 shows the mains position of a TT installation. It is necessary to install an earth electrode and it is recommended that the impedance to true earth of the electrode should not exceed 200 ohms. This can be checked by an earth loop impedance test when the supply has been connected.

Metal gas pipes or metal water pipes or other metal service pipes are not to be used as the earth electrode. A separate electrode must be installed. However, metal gas pipes, metal water pipes and other metal service pipes are required to be main bonded to the main earthing terminal as shown in Figure 3.2.3.

Because there is no protection by automatic disconnection of supply for faults on the supply side of the RCBOs the metal consumer unit must be of class II construction. Manufacturers will provide advice on the suitability of their equipment.

▼ **Figure 3.2.3**   No earth provided (TT system). Based on 25 mm² tails and selection from Table 54.7 of BS 7671.

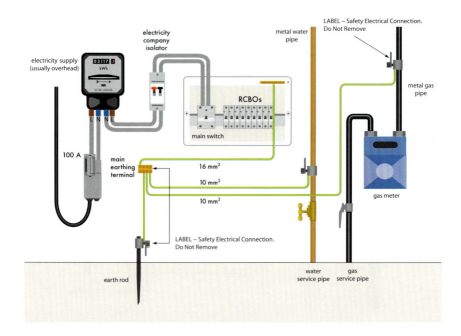

NOTE: An isolator is not always installed by the electricity distributor.

3.3   Earthing and protective equipotential bonding

3.3.1   Provision of an earthing terminal

Unless inappropriate for reasons of safety, an electricity distributor is required when providing a new connection at low voltage (230 V/400 V) to make available an earthing terminal.

For new installations, the distributor will almost always provide a PME supply.

For dwellings, it would normally always be appropriate for an earthing terminal to be provided. For farms and swimming pools, a PME supply may not be appropriate, and an installation forming part of a TT system should be employed.

3.3.2 Earthing conductor and main protective bonding conductors

Main protective bonding of metal services (Figures 3.2.1, 3.2.2, 3.2.3)

In each installation, main protective bonding conductors are required to connect to the main earthing terminal, extraneous-conductive-parts, including:

- **(a)** water service pipes;
- **(b)** gas installation pipes;
- **(c)** other service pipes (e.g. oil) and ducting;
- **(d)** central heating and air conditioning systems;
- **(e)** exposed metallic structural parts of the building; and
- **(f)** lightning protection systems (where required by BS EN 62305).

Where an installation serves more than one building, the above requirement must be applied to each building.

Plastic supply pipes

There is no requirement to main bond an incoming service where the incoming service pipe is plastic, for example, where yellow is used for natural gas and blue for potable water.

Where there is a plastic incoming service and a metal installation within the premises, main bonding is recommended unless it has been confirmed that any metallic pipework within the building is not introducing Earth potential (see section 4.3 of OSG).

All main bonding connections are to be applied to the consumer's side of any meter, main stop valve or insulating insert and, where practicable, within 600 mm of the meter outlet union or point of entry to the building if the meter is external.

Earthing conductor and main protective bonding conductor cross-sectional areas

The minimum cross-sectional area (csa) of the earthing conductor and main protective bonding conductors is given in Table 3.3.2a.

▼ **Table 3.3.2a** Earthing conductor and main protective bonding conductor sizes (copper equivalent) for TN-S and TN-C-S supplies

Line conductor or neutral conductor of PME supplies	mm²	4	6	10	16	25	35	50	70
Earthing conductor not buried or buried and protected against corrosion and mechanical damage – see notes	mm²	6	6	10	16	16	16	25	35
Main protective bonding conductor – see notes	mm²	6	6	6	10	10	10	16	25
Main protective bonding conductor for PME supplies (TN-C-S)	mm²	10	10	10	10	10	10	16	25

Notes to Table 3.3.2a:

(a) Protective conductors (including earthing and bonding conductors) of 10 mm² cross-sectional area or less shall be copper.

(b) The distributor may require a minimum size of earthing conductor at the origin of the supply of 16 mm² copper or greater for TN-S and TN-C-S supplies.

(c) Buried earthing conductors must be at least:

   (i) 25 mm² copper if not protected against mechanical damage or corrosion

   (ii) 50 mm² steel if not protected against mechanical damage or corrosion

   (iii) 16 mm² copper if not protected against mechanical damage but protected against corrosion

   (iv) 16 mm² coated steel if not protected against mechanical damage but protected against corrosion.

(d) The distributor should be consulted when in doubt.

▼ **Table 3.3.2b** Copper earthing conductor cross-sectional area (csa) for TT supplies

Buried			Not buried		
Unprotected	Protected against corrosion	Protected against corrosion and mechanical damage	Unprotected	Protected against corrosion	Protected against corrosion and mechanical damage
mm²	mm²	mm²	mm²	mm²	mm²
25	16	2.5	4	4	2.5

NOTES:

(a) Assuming protected against corrosion by a sheath.

(b) The main protective bonding conductors shall have a cross-sectional area of not less than half that required for the earthing conductor and not less than 6 mm².

Note that:

(a) only copper conductors should be used; copper-covered aluminium conductors or aluminium conductors or structural steel can only be used if special precautions outside the scope of this Guide are taken.

(b) bonding connections to incoming metal services should be as near as possible to the point of entry of the services to the premises, but on the consumer's side of any insulating section.

(c) the connection to the gas, water, oil, etc. service should be within 600 mm of the service meter, or at the point of entry to the building if the service meter is external, and must be on the consumer's side before any branch pipework and after any insulating section in the service. The connection must be made to hard pipework, not to soft or flexible meter connections.

(d) the connection must be made using clamps (to BS 951) which will not be subject to corrosion at the point of contact.

(e) if incoming gas and water services are of plastic, main bonding connections should be made to metal installation pipes only.

Earthing

Every exposed-conductive-part (a conductive part of equipment which can be touched and which is not normally live, but which can become live under earth fault conditions) must be connected to the Main Earthing Terminal.

3.3.3   Supplementary bonding conductors

In certain special locations and in installations and locations of increased shock risk, supplementary bonding is required, see Chapter 5. The cross-sectional areas of supplementary bonding conductors must comply with Table 3.3.3.

▶ **Table 3.3.3** Supplementary bonding conductor sizes

Size of protective conductor (mm²)	Minimum cross-sectional area of supplementary bonding conductors (mm²)					
	Exposed-conductive-part to extraneous-conductive-part		Exposed-conductive-part to exposed-conductive-part		Extraneous-conductive-part to extraneous-conductive-part*	
	mechanically protected	not mechanically protected	mechanically protected	not mechanically protected	mechanically protected	not mechanically protected
	1	2	3	4	5	6
1.0	1.0	4.0	1.0	4.0	2.5	4.0
1.5	1.0	4.0	1.5	4.0	2.5	4.0
2.5	1.5	4.0	2.5	4.0	2.5	4.0
4.0	2.5	4.0	4.0	4.0	2.5	4.0
6.0	4.0	4.0	6.0	6.0	2.5	4.0
10.0	6.0	6.0	10.0	10.0	2.5	4.0
16.0	10.0	10.0	16.0	16.0	2.5	4.0

* If one of the extraneous-conductive-parts is connected to an exposed-conductive-part, the bonding conductor must be no smaller than that required by column 1 or 2.

3.4   Labelling

3.4.1   Earthing and bonding connections

Bonding connections to metal pipes are made with earthing clamps to BS 951, complete with label as follows:

Durable labels as above are required to be permanently fixed in a visible position at or near:

(a) the point of connection of every earthing conductor to an earth electrode;
(b) the point of connection of every bonding conductor to an extraneous-conductive-part; and
(c) the Main Earthing Terminal where it is separate from the main switchgear.

3.4.2   Switchgear and controlgear

Unless there is no possibility of confusion, a label indicating the purpose of each item of switchgear and controlgear must be fixed on or adjacent to the switchgear or controlgear. It may be necessary to label the item controlled as well as the controlgear.

3.4.3   Distribution boards (including consumer units)

Each protective device, e.g. fuse or circuit-breaker, must be arranged and identified so that the circuit protected by the device can easily be identified.

3.4.4   Isolators

Switches used as isolators, as well as being clearly identified, must also indicate the circuit or circuits that they switch.

3.4.5   Equipment supplied from more than one source

Certain equipment may require the operation of more than one switch in order to make it safe. In such cases durable warning notices must be permanently fixed in a clearly visible position to identify the appropriate devices.

3.4.6 Periodic inspection and testing

A notice of durable material indelibly marked with the words as follows must be fixed in a prominent position at or near the origin of the installation:

> # IMPORTANT
>
> This installation should be periodically inspected and tested and a report on its condition obtained, as prescribed in the IET Wiring Regulations BS 7671 Requirements for Electrical Installations.
>
> Date of last inspection ....
>
> Recommended date of next inspection ....

3.4.7 Diagrams

A diagram, chart or schedule must be provided showing:

(a) the number of points, size and type of cable for each circuit;

(b) the method used to provide fault protection;

(c) the information necessary for the identification of each device performing the functions of protection, isolation and switching and its location; and

(d) any circuit vulnerable to an insulation test.

Circuits that may be vulnerable to insulation tests would be those that have solid-state devices, such as lighting controls, burglar alarms or fire alarms, central heating controllers, solid-state transformers, etc.

3.4.8 Residual current devices (RCDs)

Where an installation incorporates an RCD, a notice must be fixed in a prominent position at or near the origin of the installation as follows:

> This installation, or part of it, is protected by a device which automatically switches off the power supply if an earth fault develops. **Test quarterly** by pressing the button marked **'T'** or **'Test'.** The device should switch off the supply and should be then switched on to restore the supply. If the device does not switch off the supply when the button is pressed seek expert advice.

3.4.9   Warning notice: non-standard colours

If additions or alterations are made to an installation so that some of the wiring complies with the harmonized colours and there is also wiring in the old colours, a warning notice must be affixed at or near the appropriate distribution board with the following wording:

> # CAUTION
> This installation has wiring colours to
> two versions of BS 7671.
> Great care should be taken before
> undertaking extension, alteration or repair
> that all conductors are correctly identified.

3.4.10 Unexpected presence of nominal voltage exceeding 230 V

Where the nominal voltage exceeds 230 V to earth and where the presence of such a voltage would not normally be expected, a warning label stating the maximum voltage present must be provided where it can be seen before gaining access to live parts.

3.4.11 Warning notice – alternative supplies

Where an installation includes alternative supplies, such as a PV installation, which is used as an additional source of supply in parallel with another source, normally the distributor's supply, warning notices must be affixed at the following locations in the installation:

(a) at the origin of the installation;
(b) at the meter position, if remote from the origin;
(c) at the consumer unit or distribution board to which the additional or alternative supply is connected; and
(d) at all points of isolation of all sources of supply.

The warning notice must have the following wording:

> # WARNING
> ## MULTIPLE SUPPLIES
> ISOLATE ALL ELECTRICAL SUPPLIES
> BEFORE CARRYING OUT WORK
> ISOLATE MAINS AT
> ISOLATE ALTERNATIVE SUPPLIES AT

3.4.12 Warning notice – photovoltaic systems

All junction boxes (PV generator and PV array boxes) must carry a warning label indicating that parts inside the boxes may still be live after isolation from the PV convertor with the following wording:

WARNING
PV SYSTEM
Parts inside this box or enclosure may still
be live after isolation from the supply.

3.5 Installing residual current devices (RCDs)

3.5.1 Protection by an RCD

RCDs or RCBOs are required:

- **(a)** where the earth fault loop impedance is too high to provide the required disconnection time, e.g. where the distributor does not provide an earth–TT system;
- **(b)** for socket-outlet circuits (where the socket-outlet rating does not exceed 20 A);
- **(c)** for bathroom and shower-room circuits;
- **(d)** for circuits passing through zones 1 and 2 of bathrooms or shower rooms (but not necessarily supplying equipment within the zones);
- **(e)** for circuits supplying mobile equipment for use outdoors;
- **(f)** for cables without earthed metallic covering installed in walls or partitions at a depth of less than 50 mm and not protected by earthed steel conduit or similar; and
- **(g)** for cables without earthed metallic covering installed in walls or partitions with metal parts (not including screws or nails) and not protected by earthed steel conduit or the like.

30 mA RCDs are required for **(b)** to **(g)** above except for SELV and PELV circuits.

3.5.2 Applications of RCDs

Installations are required to be divided into circuits to avoid danger and minimise inconvenience in the event of a fault, and to take account of hazards that might arise from the failure of a single circuit, e.g. a lighting circuit.

The use of RCBOs, see Figure 3.5.2a, will minimise inconvenience in the event of a fault, and is applicable to all situations.

▼ Figure 3.5.2a Consumer unit with RCBOs, suitable for all installations (TN and TT)

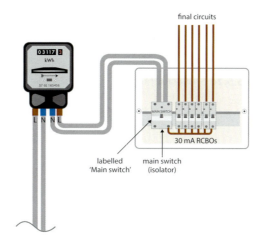

Single RCBOs protect each outgoing circuit and the risk of the busbar (connecting the supply side of each RCBO) becoming loose and making contact with the ferrous enclosure is minimal. The use of RCBOs will minimise inconvenience in the event of a fault and is applicable to all systems.

▼ Figure 3.5.2b Split consumer unit with one 30 mA RCD, suitable for TN installations with cables in walls or partitions having an earthed metallic covering or enclosed in earthed steel conduit or the like

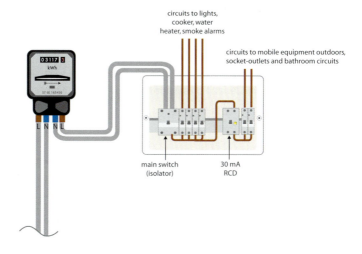

For TT installations, all circuits must be RCD protected (30 mA or 100 mA as appropriate). For cables in walls or partitions with an earthed metallic covering or installed in earthed steel conduits, 30 mA RCDs or RCBOs will only be required for bathroom circuits, socket-outlet circuits and mobile equipment outdoors and the rest of the installation needs protecting by a 100 mA RCD (see Figure 3.5.2c).

▼ **Figure 3.5.2c** Split consumer unit with time-delayed RCD as main switch, suitable for TT and TN installations with cables in walls or partitions having an earthed metallic covering or enclosed in earthed steel conduit or the like

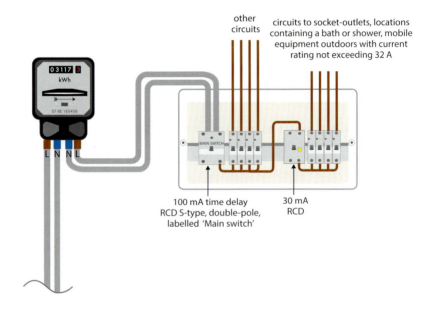

This example is suitable for installations forming part of a TT system as the time-delayed RCD will provide fault protection for the single insulated conductors supplying the 30 mA RCD.

▼ Figure 3.5.2d Split consumer unit with separate main switch and two 30 mA RCDs, suitable for all installations

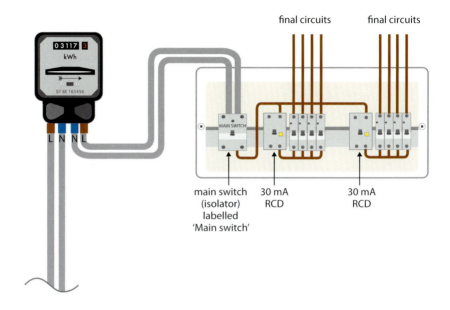

final circuits   final circuits

main switch   30 mA   30 mA
(isolator)   RCD   RCD
labelled
'Main switch'

The division of an installation into two parts with separate 30 mA RCDs will ensure that part of the installation will remain on supply in the event of a fault, see Figure 3.5.2d. Generally, this is not suitable for an installation forming part of a TT system as there is insufficient fault protection of the single insulated conductors which connect the load side of the double-pole main switch to the supply side of the RCCBs.

NOTE: Residual current device (RCD) is a device type that includes residual current circuit-breakers (RCCBs), residual current circuit-breakers with integral overcurrent protection (RCBOs) and socket-outlets incorporating RCDs (SRCDs).

Circuit specifications

4

NOTE: The calculation of load current from kVA and kW is described in sections 9.6 to 9.9.

4.1 Standard circuits

The standard circuits have been designed for dwellings where the supply is at 230 V single-phase or 400 V three-phase.

The supply conditions assumed are those typical for the UK:

(a) the installation is supplied by:
    (i) a TN-C-S supply with a maximum external earth fault loop impedance, Z_e, of 0.35 ohm; or
    (ii) a TN-S supply with a maximum Z_e of 0.8 ohm; or
    (iii) a TT supply.

(b) each final circuit is connected to a distribution board or consumer unit at the origin of the installation.

(c) the method of installation complies with reference methods C, B, 102, 100, A (see section 2.3). Increased conductor sizes only for reference method 101 are also given in the tables in this section.

(d) the ambient temperature throughout the length of the circuit does not exceed 30 °C.

(e) circuit lengths are 'rounded', recognising the impracticality of exact measurement.

(f) cables may be thermosetting (LSHF) or thermoplastic (PVC).

4.1.1 Grouping of circuit cables

For cables of household or similar installations, with the exception of heating and water heating, if the rules below are followed, increased cable sizes to allow for grouping are not usually necessary:

(a) cables are not grouped when installed in or under insulation, that is, for installation methods 100, 101, 102 or 103;

(b) cables clipped direct (including on cement or plaster) are clipped side by side in one layer preferably separated by one cable diameter; and

(c) cables above ceilings are clipped to joists as per installation reference methods 100 and 101 of Table 4A2 of BS 7671, see Figure 2.3.1.

Whole house heating and water heating cables are not to be grouped unless cables with a suitably increased cable cross-sectional area are installed.

Manufacturer's advice on the suitability of the distribution board or consumer unit should also be sought for electric whole house heating.

4.1.2 White and grey sheathed cables

Insulated and sheathed flat cables with bare protective conductor in the new harmonized cable colours are available in two types as follows:

▶ **White sheathed** – thermosetting cable with low smoke (LSHF) properties.

▶ **Grey sheathed** – PVC cable as previously available in the old cable colours.

Insulated and sheathed flat cables with bare protective conductor in the old cable colours may have white or grey sheaths, but they are always PVC cables.

▼ **Table 4.1.2a** Type B standard domestic circuits with grey thermoplastic (PVC) or white thermosetting cable, installation methods A, 100, 102, B and C. Type B circuit-breaker to BS EN 60898, Type B RCBO to BS EN 61009 and Type 1 mcb to BS 3871 (for existing installations)

Type of final circuit	Cable csa PVC/ PVC	Circuit-breaker rating	Maximum cable length				Maximum test loop impedance[1]	Minimum csa for Method 101
			TN-C-S PME earth Z_e up to 0.35 Ω		TN-S sheath earth Z_e up to 0.8 Ω			
				RCD		RCD		
	mm^2	A	m	m	m	m		mm^2
Ring, supplying 13 A socket-outlets	2.5/1.5	32	NP	100	NP	100	1.1	4.0

Type of final circuit	Cable csa PVC/ PVC	Circuit-breaker rating	Maximum cable length				Maximum test loop impedance[1]	Minimum csa for Method 101
			TN-C-S PME earth Z_e up to 0.35 Ω		TN-S sheath earth Z_e up to 0.8 Ω			
				RCD		RCD		
	mm^2	A	m	m	m	m		mm^2
Radial, supplying 13 A socket-outlets	2.5/1.5	20	NP	40	NP	40	1.75	4.0
Cooker (oven + hob + socket-outlet)	6/2.5	32	50	50	50	50	1.1	10.0
Oven (no hob)	2.5/1.5	16	50	50	50	50	2.2	2.5
Immersion heater	2.5/1.5	16	50	50	50	50	2.2	2.5
Shower to 30 A (7.2 kW)	6/2.5	32	NP	50	NP	50	1.1	10.0
Shower to 40 A (9.6 kW)	10/4	40	NP	65	NP	65	0.88	16.0
Storage radiator	2.5/1.5	16	50	50	50	50	2.2	2.5
Fixed lighting[3,4]	1.5/1.0	6	100	100	100	100	5.87	1.5
Fixed lighting[3,4]	1.5/1.0	10	85	100	75	100	3.50	1.5
Fixed lighting in locations with a bath or shower[4]	1.5/1.0	10	NP	100	NP	100	3.50	1.5

NOTES:
1 Measured values at an ambient temperature of 10 °C and above.
2 Cable cross-sectional area will usually need to be increased for installation reference method 101, see Figure 2.3.1g.
3 For loop in systems, excludes intermediate switch drops.
4 Distributed load, 2-5 A at extremity.
NP Not Permitted as RCD is required.

▼ **Table 4.1.2b** Type C standard domestic circuits grey thermoplastic (PVC) or white thermosetting cable, installation methods A, 100, 102, B and C. Type C circuit-breaker to BS EN 60898, Type C RCBO to BS EN 61009 and Type 3 mcb to BS 3871-1 (for existing installations)

Type of final circuit	Cable csa PVC/ PVC	Circuit-breaker rating	Maximum cable length				Maximum test loop imped-ance[1]	Minimum csa for Method 101
			TN-C-S PME earth Z_e up to 0.35 Ω	TN-S sheath earth Z_e up to 0.8 Ω				
				RCD		RCD		
	mm²	A	m	m	m	m	Ω	mm²
Ring, supplying 13 A socket-outlets	2.5/1.5	32	NP	100	NP	100	0.55	4.0
Radial, supplying 13 A socket-outlets	2.5/1.5	20	NP	40	NP	40	0.87	4.0
Cooker (oven + hob + socket-outlet)	6/2.5	32	NP	50	NP	50	0.55	10.0
Oven (no hob)	2.5/1.5	16	30	50	12	50	1.09	2.5
Immersion heater	2.5/1.5	16	30	50	12	50	1.09	2.5
Shower to 30 A (7.2 kW)	6/2.5	32	NP	50	NP	50	0.55	10.0
Shower to 40 A (9.6 kW)	10/4	40	NP	65	NP	65	0.44	16.0
Storage radiator	2.5/1.5	16	30	50	12	50	1.09	2.5
Fixed lighting[3,4]	1.5/1.0	6	70	100	55	100	2.91	1.5
Fixed lighting[3,4]	1.5/1.0	10	35	100	25	100	1.75	1.5
Fixed lighting in locations with a bath or shower[4]	1.5/1.0	10	NP	100	NP	100	1.75	1.5

NOTES:
1 Measured values at an ambient temperature of 10 °C and above.
2 Cable cross-sectional area will usually need to be increased for installation reference method 101, see Figure 2.3.1g.
3 For loop in systems excludes intermediate switch drops.
4 Distribution load, 2.5 A at extremity.
NP Not Permitted as RCD is required.

Table 4.1.2c BS 88-3 standard domestic circuits grey thermoplastic (PVC) or white thermosetting cable, installation methods A, 100, 102, B and C. Cartridge fuse to BS 1361 for existing circuits

Type of final circuit	Cable csa PVC/PVC	Fuse rating	Maximum cable length				Maximum test loop impedance[1]	Minimum csa for Method 101
			TN-C-S PME earth Z_e up to 0.35 Ω		TN-S sheath earth Z up to 0.8 Ω			
				RCD		RCD		
	mm²	A	m	m	m	m	Ω	mm²
Ring, supplying 13 A socket-outlets	2.5/1.5	32	NP	100	NP	100	0.73	4.0
Radial, supplying 13 A socket-outlets	2.5/1.5	20	NP	40	NP	40	1.55	4.0
Cooker (oven + hob + socket-outlet)	6/2.5	32	NP	50	NP	50	0.73	10.0
Immersion heater	2.5/1.5	16	50	50	40	50	1.84	2.5
Shower to 30 A (7.2 kW)	6/2.5	32	NP	50	NP	50	0.73	10.0
Shower to 40 A (9.6 kW)	10/4	45	NP	65	NP	65	0.76	16.0
Storage radiator	2.5/1.5	16	50	50	40	50	1.84	2.5
Fixed lighting[3,4]	1.5/1.0	5	100	100	100	100	7.9	1.5
Fixed lighting in locations with a bath or shower[4]	1.5/1.0	5	NP	100	NP	100	7.9	1.5

NOTES:
1 Measured values at an ambient temperature of 10 °C and above, no RCD.
2 Cable cross-sectional area will usually need to be increased for installation reference method 101, see Figure 2.3.1g.
3 For loop in systems, excludes intermediate switch drops.
4 Distributed load, 2.5 A at extremity.
NP Not Permitted as RCD is required.

4.2 Final circuits using 13 A socket-outlets (to BS 1363-2) and fused connection units (to BS 1363-4)

4.2.1 General

In dwellings, all socket-outlets other than specifically labelled sockets are required to be protected by a 30 mA RCD.

Ring circuits should not supply immersion heaters, comprehensive space heating, ovens and hobs with a load exceeding 2 kW or similar loads. Socket-outlets should be installed to provide for a reasonable spread of the load around the ring.

▼ **Table 4.2.1** Minimum conductor sizes and maximum floor area for socket-outlet circuits in household and similar premises

	Overcurrent protective device rating (A)	Minimum conductor cross-sectional area for copper conductor thermoplastic (PVC) or thermosetting insulated cables (mm²)	Maximum floor area supplied (m²)
Ring circuit	30 or 32	2.5	100
Radial circuit	20	2.5	50
Radial circuit	30 or 32	4.0	75

A ring or radial socket-outlet final circuit, with spurs and permanently connected equipment, if any, may supply an unlimited number of socket-outlets and fused connection units within the area given in Table 4.2.1.

Socket-outlets for washing machines, tumble dryers and dishwashers should be located so as to provide reasonable sharing of the load in each leg of the ring.

Where two or more ring final circuits are installed, the socket-outlets and permanently connected equipment to be served are to be reasonably distributed among the circuits.

4.2.2 Spurs

A non-fused spur feeds only one single or one twin or multiple socket-outlet or one permanently connected item of equipment. Such a spur is connected to a circuit at the terminals of a socket-outlet or junction box or at the origin of the circuit in the distribution board (Figure 4.2.2).

A non-fused spur is wired in the same size (cross-sectional area) cable as the ring final circuit. A fused spur is connected to the circuit through a fused connection unit, the rating of the fuse in the unit not exceeding that of the cable forming the spur and, in any event, not exceeding 13 A.

▼ **Figure 4.2.2** Ring final circuit supplying socket-outlets and an item of
permanently connected equipment

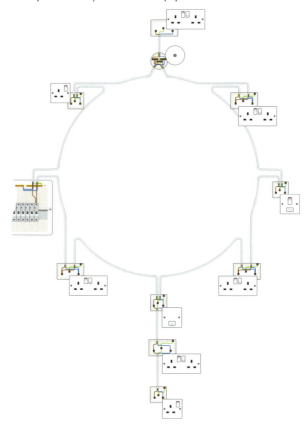

4.3 Permanently connected equipment: requirements for switching

Equipment permanently connected to a socket-outlet circuit and not exceeding 13 A rating should be locally protected by a fuse of rating not exceeding 13 A (to BS 1362) or by a circuit-breaker of rating not exceeding 16 A and controlled by a readily accessible switch in the fixed installation (e.g. switched fused connection unit). A separate switch is not required where a local circuit-breaker is used.

Each switch or circuit-breaker used for isolation must be identified by position or durable marking to indicate the equipment controlled.

4.4 Cooker circuits in household or similar premises

A cooker circuit with cables in walls or partitions installed in earthed steel conduit or with an earthed metallic sheath does not require RCD protection unless the cooker control unit incorporates a socket-outlet.

The circuit should supply a control switch complying with BS 3676 (BS EN 60669-1) or a cooker control unit complying with BS 4177.

The rating of the circuit is determined by the assessment of the current demand of the cooking appliance(s) and the cooker control unit socket-outlet, if any. A 30 A or 32 A circuit is usually appropriate for household or similar cookers of rating up to 15 kW.

A circuit of rating exceeding 15 A but not exceeding 50 A may supply two or more cooking appliances where these are installed in one room (Figure 4.4). It is recommended that a control switch or cooker control unit should be installed and placed within two metres of the appliance, but not directly above it. Where two stationary cooking appliances are installed in one room, one switch may be used to control both appliances provided that neither appliance is more than two metres from the switch.

It is recommended that electric ovens are supplied by a separate circuit; however, for lightly loaded circuits, ovens of rating 13 A or less may be connected into a ring final circuit.

Precautions need to be taken to prevent the heat generated by a cooker, including the hob, creating a risk of fire, or of harmful thermal effects, to adjacent equipment or materials including furnishings, e.g. curtains.

See section 5.2 for location of accessories in kitchens.

▼ **Figure 4.4** Cooker circuit

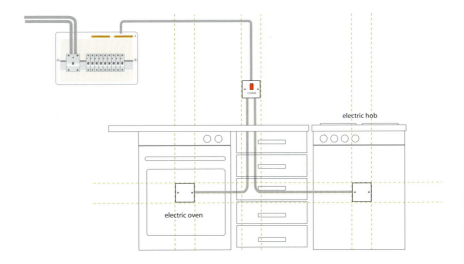

NOTE: Cables without an earthed metallic covering to be installed at a depth greater than 50 mm or enclosed in earthed steel conduit or protected by 30 mA RCD.

4.5   Water and space heating

A water heater fitted to a storage vessel in excess of 15 litres capacity, or a permanently connected heating appliance forming part of a space heating installation, should be supplied by its own separate circuit (not from a ring final circuit).

Immersion heaters should be supplied through a switched flex-outlet connection unit (to BS 1363-4) or a double-pole switch with flex outlet complying with BS EN 60669-1 or BS EN 60669-2-4.

Instantaneous water heaters (including showers) of rating exceeding 3 kW should be supplied by a separate circuit (see section 4.1). Local isolation of water and electricity should be provided to facilitate maintenance.

Instantaneous water heaters including showers must comply with BS EN 60335-2-35. Check they are water-protected to a minimum IPX4 rating. Shower pumps must comply with BS EN 60335-2-41 and once again should be protected to a minimum IPX4 rating.

4.6   Lighting circuits

See the tables in this chapter for specifications for domestic circuits, Figure 4.6 for circuits wired in the harmonized colours and also Figures 11.3 and 11.4.

Where lighting circuits are installed for, or passing through, locations containing a bath or a shower 30 mA RCD protection must be provided.

▼ **Figure 4.6** Domestic lighting circuit

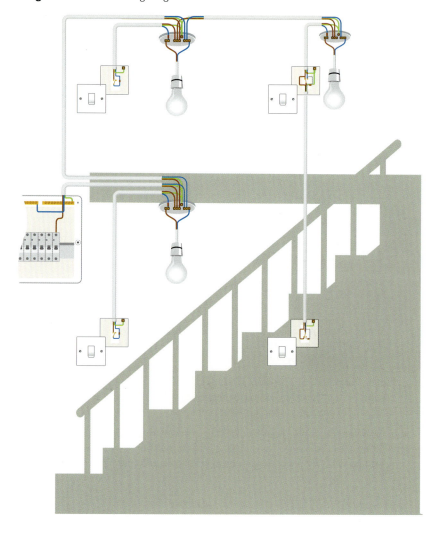

Kitchens, special installations, special locations, etc.

5

5.1 Introduction

The Wiring Regulations BS 7671

Special installations or locations in BS 7671 terminology are those installations or locations where there are additional requirements. The additional requirements, for example, might be to protect against environmental conditions, such as water and steam, and to provide adequate protection against electric shock in situations where the body is more susceptible, i.e. without clothes, immersed in water or in contact with the general mass of Earth.

Part P of the Building Regulations

In Approved Document P, certain of the special locations of BS 7671 are identified as locations where the relaxation allowing minor works* to be carried out without notifying Building Control does not apply. These locations are:

- within zones 0, 1 and 2 of a room with a bath or shower;
- in a room with a swimming pool; and
- in a room with a sauna heater.

Only the replacement of equipment including accessories can be carried out without notification in these locations.

*Minor work does *not* include:

- installing a new circuit; or
- replacing a consumer unit.

5.2 Kitchens

5.2.1 Introduction

A kitchen is no longer identified in the Approved Document as being a special location. However, the work in installing a fitted kitchen can be fairly substantial, involving plumbing, ventilation and significant alterations to the electrical installation, so it is worthwhile to provide some guidance.

5.2.2 Location of accessories in kitchens

General guidance can be provided as follows:

(a) Wiring accessories (for example, socket-outlets, switches) should be permanently and securely fixed and readily accessible. Accessories and electrical equipment such as socket-outlets and under-cupboard lighting can be fixed to fitted kitchen units provided that they are securely fixed to rigid parts of the units that are not demountable or otherwise liable to be disturbed in normal service. Account should be made for the accessibility for inspection, testing and maintenance, and provision of adequate protection against damage (for example, by impact or water) for the accessories, equipment and associated wiring.

(b) Cooker control switches, extractor fan switches, etc. should not be mounted so that it is necessary to lean or reach over gas or electric hobs for their operation.

(c) Socket-outlets should be installed a minimum of 450 mm from the floor.

(d) Accessories should be installed a minimum of 300 mm from the edge of kitchen sinks and draining boards to reduce the risk of being splashed.

(e) Socket-outlets supplying washing machines and dishwashers, etc. should be positioned so that water that may drip from plumbing or the equipment is unlikely to affect the socket-outlet or plug.

(f) To prevent damage to the plug and flexible cable on insertion and withdrawal the centre of a socket-outlet should be a minimum of 150 mm above the work surface.

(g) Socket-outlets supplying appliances pushed under a work surface, e.g. dishwashers, tumble dryers and fridges, should be accessible when the appliance is pulled out.

(h) Appliances built into kitchen furniture (integrated appliances) should be connected to a socket-outlet or fused connection unit that is readily accessible when the appliance is in place and in normal use, or be supplied from a socket-outlet or other connecting device controlled by a readily accessible double-pole switch or switched fused connection unit.

(i) Light switches should be readily accessible.

(j) Cooker hoods should be 650 mm to 700 mm above the hob surface, subject to (k); see section 10.5 for extract requirements.

(k) Installers shall take account of manufacturer's instructions.

▼ **Figure 5.2.2** Kitchen installation

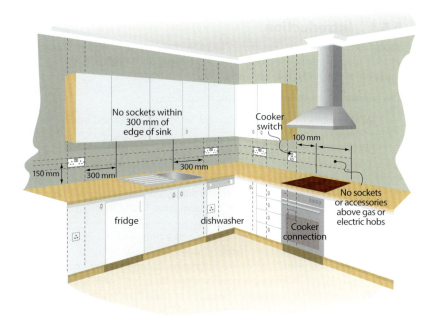

NOTES:

(a) Cables without an earthed metallic covering to be installed at a depth of greater than 50 mm or enclosed in earthed steel conduit, or protected by a 30 mA RCD.

(b) Metal waste pipes in contact with Earth may be extraneous-conductive-parts and should be main bonded back to the main earthing terminal.

(c) Socket-outlets must be protected by a 30 mA RCD.

5.2.3   Supplementary bonding in kitchens

There is no specific requirement in BS 7671 to provide supplementary bonding in kitchens. Water pipes, metal sinks or draining boards and metal furniture do not require supplementary bonding.

5

5.3   Locations containing a bath or shower

NOTE: Circuits supplying equipment in bath and shower locations must be protected by a 30 mA RCD.

5.3.1   Introduction

Special precautions are required in locations containing a bath or shower, including shower rooms, en-suite showers and bedrooms with a bath or a shower. Bedrooms with a shower enclosure are covered in section 5.3.2.

The additional requirements are summarised below:

(a) All circuits of the location (except SELV and PELV) must be 30 mA RCD protected.

(b) All circuits (except SELV and PELV) passing through zones 1 and 2 of the location, including those not supplying equipment in the location, must be 30 mA RCD protected.

(c) Socket-outlets, other than shaver supply units complying with BS EN 61558-2-5, are not allowed within 3 m from zone 1 (and must be protected by a 30 mA RCD).

(d) Equipment installed in zones 0, 1 and 2 must have suitable additional protection against the ingress of water, see Table 5.3.1.

(e) There are restrictions on the installation of current-using equipment, switchgear and wiring accessories, see Table 5.3.1 and Figure 5.3.1a.

(f) Except for SELV systems, the cables of underfloor heating installations must be covered by an earthed metallic grid or the heating cable should have an earthed metallic sheath.

▼ **Table 5.3.1** Requirements for equipment (current-using and accessories) in locations containing a bath or shower

Zone*	Minimum degree of protection	Current-using equipment (appliances)	Switchgear and accessories
0	IPX7	Only 12 V a.c. and 30 V d.c. SELV fixed equipment, the safety source being installed outside the zones.	None allowed.
1	IPX4	25 V a.c. and 60 V d.c. SELV or PELV equipment allowed, the safety source being installed outside the zones. The following fixed, permanently connected equipment are allowed: whirlpool units, electric showers, shower pumps, ventilation equipment, towel rails, luminaires, water heaters.	Only 12 V a.c. and 30 V d.c. switches of SELV circuits allowed, the source being installed outside zones 0, 1 and 2.
2	IPX4 (but shaver supply units complying with BS EN 61558-2-5 allowed when protected from direct shower spray)	Fixed, permanently connected equipment allowed.	SELV switches and sockets allowed, the source being outside zones 0, 1 and 2, and shaver supply units to BS EN 61558-2-5 allowed only if fixed where direct spray from showers is unlikely.
Outside	General requirements	General requirements.	Accessories allowed and SELV sockets and shaver supply units to BS EN 61558-2-5. Socket-outlets allowed 3 m from the edge of zone 1.

* Protection by 30 mA RCD is required for all low voltage circuits to locations containing a bath or shower.

5

▼ **Figure 5.3.1a** Elevation of bathroom zones

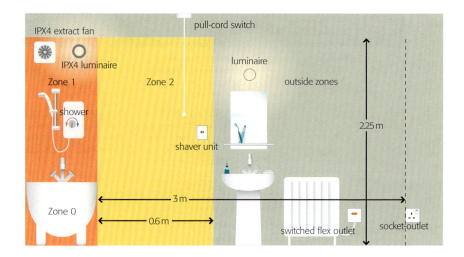

▼ **Figure 5.3.1b** Plan of bath tub with partition

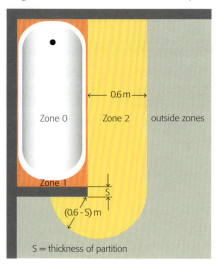

▼ **Figure 5.3.1c** Plan of shower with partition and shower tray

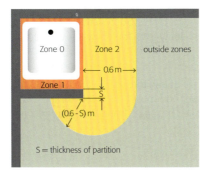

▼ **Figure 5.3.1d** Plan of shower with partition but without shower tray

S = thickness of partition

Y = radial distance from the fixed water outlet to the inner corner of the partition

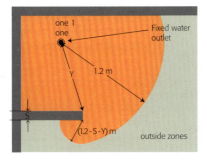

Socket-outlets

13 A socket-outlets are allowed only if at least 3 m horizontally from the boundary of zone 1. Shaver supply units complying with BS EN 61558-2-5 are allowed in zone 2 and outside the zones, if installed where direct spray from the shower is unlikely.

Plate switches

A plate switch is allowed outside the zones of a bathroom. A switch should be at least 0.6 m from the edge of the bath or shower and must be suitable for the location. The cords of cord-operated switches are allowed in zones 1 and 2 and are recommended for bathrooms and shower rooms.

Luminaires (light fittings)

230 V fittings may be installed above a shower or bath but they must be at least IPX4 rating, i.e. enclosed and water protected. If installed more than 0.6 m from the edge of a shower basin or bath, no special fitting is required but the luminaire must be of a suitable design for the conditions.

Shower units

Electric showers and electric shower pumps should comply with BS EN 60335-2-35 and BS EN 60335-2-41 respectively. Such showers are usually suitable for installation within zone 1. It is normal practice to provide an isolation switch within the bathroom. The switch must be installed outside zones 0, 1 and 2 although the cord of cord-operated switches may reach into zones 1 or 2.

Extractor fans

A suitable 230 V extractor fan may be installed in zones 1 and 2, and outside the zones. If an extractor fan is installed in zone 1 or 2 it must be protected against the ingress of moisture to at least IPX4 rating.

An extractor fan supplied from a lighting circuit for a bathroom without a window should have its own means of isolation; if it does not, the replacement or maintenance of the fan would have to be carried out in the dark. An isolation switch for a fan with an overrun facility will need to be triple-pole (switch wire, line and neutral) and must be installed outside zones 0, 1 and 2.

Equipment installed in the zones

Manufacturer's instructions must confirm that the equipment is suitable for use in the relevant zone. Equipment selection should not be based solely upon its IP rating.

Fixed equipment

Fixed equipment that has no protection against ingress of water can be installed outside the zones of bathrooms including the provision of a switched fused spur, providing such equipment is of a suitable design for the conditions.

Home laundry equipment

Washing machines and tumble dryers may be installed in a location with a bath or shower provided they are:

- ▶ supplied from a switched fused flex outlet (socket-outlets are only allowed at least 3 m from zone 1) installed outside the zones;
- ▶ protected by a 30 mA RCD; and
- ▶ permitted for such installation by the manufacturer.

Electric floor heating

Floor heating installations installed below any of the zones must be protected by SELV or have an overall earthed metallic grid or the heating cable should have an earthed metallic sheath, connected to the protective conductor of the supply circuit.

5.3.2 Bedrooms with a shower enclosure

The requirements for bathrooms (section 5.3.1) are to be met for bedrooms with a shower cubicle. Socket-outlets are allowed in bedrooms containing a shower cubicle provided they are at least 3 m from the shower cubicle. Within the location all circuits are to be protected by a 30 mA RCD. See Figure 5.3.2.

▼ **Figure 5.3.2**   Zones in a bedroom with a shower

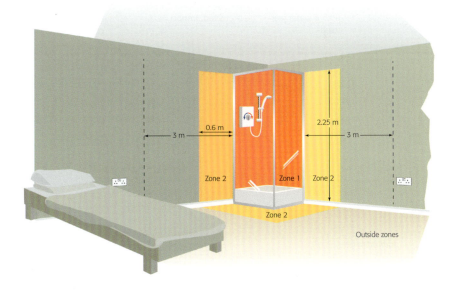

5

5.4   Swimming pools, hot tubs, etc.

5.4.1   Introduction

Persons carrying out installations in swimming pools and other basins, for example, hot tubs, must make themselves familiar with Section 702 of BS 7671. The following table provides an introduction to the requirements.

▼ **Table 5.4.1**   Swimming pools and other basins: some requirements for equipment in zones

Zone (see Fig. 5.4.1)	Minimum degree of protection 702.3512.2	Current-using equipment 702.410.3, 702.55	Switchgear, control gear accessories
0	IPX8	SELV 12 V a.c. or 30 V d.c., source outside zones or in zone 2 with RCD to source Impact AG2.	Not allowed.
1	IPX4 IPX5 if water jets used	SELV 25 V a.c. or 60 V d.c., source outside zones or in zone 2 with RCD to source Impact AG2. See 5.4.2 for pumps, heaters. See5.4.3 for luminaires.	Not allowed, see 5.4.4.
2	IPX2 indoors IPX4 outdoors IPX5 if water jets used	SELV 50 V a.c. or 120 V d.c. if source in zone 2 RCD protected. Impact AG2.	Not allowed. Sockets to be 30 mA RCD protected or SELV or electrically separated.

▼ **Figure 5.4.1** Examples of zone dimensions (plan) with fixed partition height of at least 2.5 m

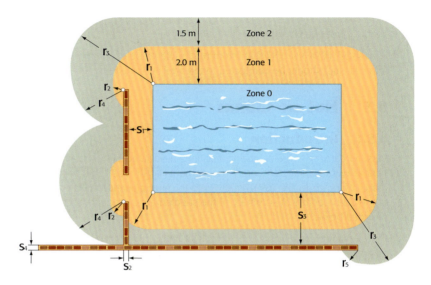

5.4.2 Pumps, heaters, filters etc.

Pumps, heaters, filters, etc. in zone 1 may be supplied at low voltage if:

▶ they are protected by 30 mA RCD or electrically separated;
▶ they are only accessible via hatch or door secured by key or tool; and
▶ the opening of the hatch or door disconnects all live conductors.

5.4.3 Luminaires

Luminaires in zone 0 must comply with BS EN 60598-2-18.

Where there is no zone 2, low voltage Class II, AG2 luminaires may be installed in zone 1 if mounted at least 2 m from the floor and protected by a 30 mA RCD.

5.4.4 Socket-outlets and switches in zone 1

If a pool is constructed such that it is not possible to install switches or socket-outlets outside zone 1, they are permitted in zone 1 if they are:

▶ of insulated construction;
▶ installed at least 1.25 m from zone 0;
▶ at least 0.3 m above the floor; and
▶ protected by 30 mA RCD, or SELV (25 V a.c. or 60 V d.c.) or electrically separated, and the switch or socket is labelled 'For use only when pool evacuated'.

5.4.5 Electricity supplies and bonding

The requirements of the electricity distributor must be complied with.

All extraneous-conductive-parts in the zones must be connected to the protective conductor of equipment in the zones.

For TN-C-S (PME) supplies, an earth mat or electrode of suitably low resistance (20 Ω for domestic) should be connected to the pool earthing terminal.

5.5 Hot air saunas

5.5.1 Introduction

Persons carrying out electrical installations associated with hot air saunas should be familiar with the supplementary requirements of BS 7671, Section 703 and need to make reference to BS 7671 directly and to other guidance such as Guidance Note 7 or the *Commentary*. A summary of the supplementary requirements in Section 703 is provided below.

Hot air saunas are divided into zones 1, 2 and 3 for the purpose of selecting suitable equipment, see Figure 5.5.4.

5.5.2 Summary of requirements

- ▶ Sauna heating appliances should meet the requirements of BS EN 60335-2-53.
- ▶ In zone 1, only the sauna heater and directly related equipment is to be installed.
- ▶ In zone 2, only luminaires and control devices for the sauna heater required to be in zone 2, and associated wiring, are allowed. Light switches are required to be outside the sauna.
- ▶ All equipment must have a degree of protection of at least IPX4 rating; if water jets are used for cleaning then IPX5 rating.
- ▶ Temperature ratings – equipment in zone 3 must be suitable for an ambient temperature of 125 °C and cables shall withstand a temperature of 170 °C.
- ▶ All circuits shall be protected by one or more 30 mA RCDs except the sauna heater circuit, unless recommended by the manufacturer.

5.5.3 Wiring

The wiring should preferably be installed on the 'cold side' of the thermal insulation. If installed on the 'hot side' in zones 1 and 3, it must be suitable for 170 °C.

5.5.4 Equipment

Only essential equipment should be installed in the sauna, i.e. equipment for supplying the sauna heater itself and a luminaire. Wherever possible, electrical equipment should be installed outside the sauna room or cabin.

▼ **Figure 5.5.4**   Temperature zones of a hot air sauna

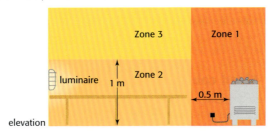

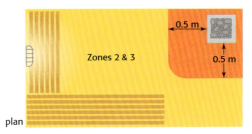

5.6   Floor and ceiling heating systems

Requirements for floor and ceiling heating systems are outside the scope of this publication and are found in Section 753 of BS 7671.

Ceiling heating systems pose the risk of penetration of the heating element by nails, pins, etc. pushed through the ceiling surface. For this reason, additional protection against electric shock is required by the use of a 30 mA RCD.

Similarly, there is concern that floor heating installations could be damaged by carpet gripper rods, nails, etc. and, once again, protection by a 30 mA RCD is required.

To protect the building structure against the risk of fire, there are particular requirements to avoid overheating of the floor or ceiling heating systems.

Persons installing such systems should obtain specialist advice. Guidance is given in Guidance Note 7: *Special Locations*.

5.7   Outdoor lighting and power, sheds, garages and greenhouses

Outdoor lighting and power and electrical installations in sheds, garages and greenhouses are within the scope of Part P of the Building Regulations.

5.7.1   The risk

The general rules for outdoor circuits and equipment apply to all gardens including domestic gardens:

▶ all socket-outlets with a current rating not exceeding 20 A are required to be protected by a 30 mA RCD; and

▶ mobile equipment (including portable equipment) with a current rating not exceeding 32 A whether supplied by a socket-outlet or a flex outlet is required to be protected by a 30 mA RCD.

5.7.2 Fixed cables

Cables must be permanently fixed in a protected location or mechanically protected or buried. It must be remembered that the layout of a garden can be changed and care should be taken to install cables where they are not likely to be damaged or disturbed, such as by laying the cable around the edge of the plot and at sufficient depth.

Cables buried in the ground must be buried at a sufficient depth to avoid damage by any disturbance of the ground that is reasonably likely to occur. Cables should generally be buried at least 500 mm, preferably deeper, below the lowest ground level and route marker tape used, laid along the cable route approximately 150 mm below the surface. Buried cables must be steel-wire armoured or metal sheathed or enclosed in a conduit or duct. Double digging is likely to occur in a vegetable plot and, if cable must be laid in such a location, it should be at least 600 mm deep.

A buried cable route should be identified by route markers and recorded on drawings retained with the Electrical Installation Certificate.

Above ground cables need to be shielded against prolonged exposure to direct sunlight, particularly grey and white PVC cables. Cables with a black or rubber sheath are recommended if direct exposure cannot be avoided. Ultraviolet light from the sun will degrade plastics and, unless shielded, will shorten the life. Black PVC and rubber cables have a reasonable life outdoors.

5.7.3 Supplies to outbuildings

When a cable is laid from a house to an outbuilding such as a shed or greenhouse and the outbuilding has extraneous-conductive-parts, a main protective bonding conductor must connect the extraneous-conductive-parts (such as metal water or gas pipes) to the main earthing terminal (in the house).

For an outbuilding fed from a dwelling with a PME supply, the bonding conductor must comply with Table 54.8 of BS 7671, that is, have a minimum csa of 10 mm² copper.

NOTE: If there are no extraneous-conductive-parts in the outbuilding the main equipotential bonding will not be required, see Figure 5.7.3. The insertion of an insulating length of pipe or use of plastic piping in the building will eliminate extraneous-conductive-parts.

▼ Figure 5.7.3 Outbuilding without extraneous-conductive-parts

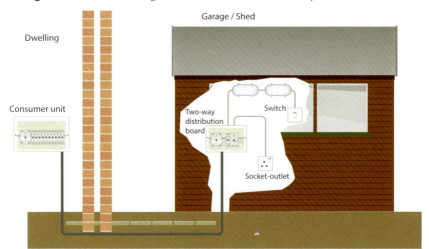

Alternatively, the outbuilding can be made part of a TT system. A local earth electrode is required at the outbuilding connected to the outbuilding earth terminal.

The earthing of a TT outbuilding must not be connected to the earthing of the main building. If an armoured cable supplies the outbuilding, care must be taken to ensure the armouring is not connected to the earth terminal of the outbuilding (it must be connected to the earthing terminal of the main building.)

5.7.4 Socket-outlets

Socket-outlets installed out of doors must be of a weatherproof construction (e.g. IP54) and must be protected by a 30 mA RCD.

5.7.5 Fixed equipment

Fixed equipment in the garden, such as permanent lighting attached to buildings, should be securely erected with all cables buried or securely fixed to permanent structures clear of the ground. All-insulated Class II equipment is recommended where possible. Outdoor fixed equipment is not required to be protected by an RCD. A disconnection time of 0.4 s is required for circuits not exceeding 32 A.

5.7.6 Ponds

Ponds are a natural source of concern because of the presence of water. All equipment must be specifically designed for pond use and consequently be of a suitable IP rating or installed in a suitable enclosure. Class II equipment is recommended, i.e. equipment with an all-insulated enclosure. Where practicable, cables should be installed in ducts or conduits built into the pond structure and not left loose on the ground. All electrical connections should be made in robust water-resistant junction boxes that have an IP rating of IP55 or better.

Pond lighting should meet the requirements of BS EN 60598-1 generally, and BS EN 60598-2-18 if in contact with, or immersed in, water. Luminaires in contact or immersed should be IPX8, if not in contact or immersed, IP54. Pumps should meet the requirements of BS EN 60335-2-41 and other equipment should meet the requirements of BS EN 60335-2-55.

5.8 Solar photovoltaic systems and small-scale embedded generators

The installation of such equipment is outside the scope of this Guide. The requirements for solar photovoltaic systems are given in Section 712 of BS 7671 and guidance is given in IET Guidance Note 7: *Special Locations.*

5.9 Builders' supplies

Hand tools including handlamps on any work site including domestic re-wires should preferably be supplied from a 110 V source or be protected by a 30 mA RCD.

For installation of supplies to construction sites, reference needs to be made to BS 7671 Section 704 and Guidance Note 7, also HSE publication HSG141 (formerly GS24) *Electrical Safety on Construction Sites.*

BS 7671 recommends 110 V single-phase centre-tapped 55 V to earth supplies for portable hand lamps and hand tools, and local lighting up to 2 kW. SELV is strongly preferred for portable hand lamps in confined or damp locations.

5.10 Electric vehicle charging installations

For PME supplies, particular precautions must be taken for electric vehicle charging installations with the charging point mounted outdoors. Guidance is given in the IET Code of Practice for Electric Vehicle Charging.

This section introduces the installer to the likely requirements for PME installations.

▼ **Figure 5.10.1** Typical domestic PME supply schematic with additional earth electrodes

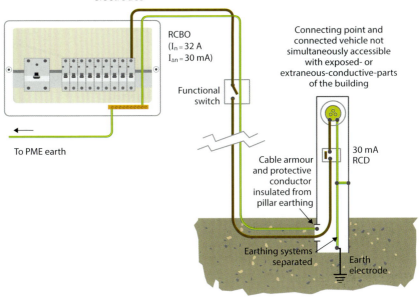

In dwellings, the demand for the electric vehicle charging equipment must not exceed 32 A unless particular arrangements have been made with the local network operator.

For PME supplies the additional requirements for an outdoor circuit include:

(a) each charging point must be protected by a type A 30 mA RCD;

(b) the lowest part of the charging point must be between 0.5 and 1.5 from the ground;

(c) a main bonding conductor must be installed to the charging point with a minimum csa of 25 mm² copper; and

(d) a minimum of two 2.4 m or three 1.2 m electrodes separated by 3 m be installed, see Figure 510.1.

For vehicle connecting points installed such that the vehicle can only be charged within a building, for example, in a garage with a (non-extended) tethered lead, see Figure 510.2. The PME earth may be used without additional earth electrodes.

▼ **Figure 5.10.2** Tethered charging cable

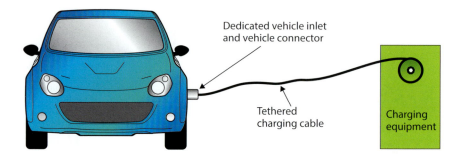

Dedicated vehicle inlet
and vehicle connector

Tethered
charging cable

Charging
equipment

Inspection and testing 6

6.1 Introduction

6.1.1 BS 7671:2008+A3:2015

Irrespective of who has carried out the installation it must be inspected and tested in accordance with the requirements of the IET Wiring Regulations (BS 7671:2008+A3:2015 *Requirements for Electrical Installations*) including inspection, testing, certification and reporting. The work is not completed until it has been inspected and tested and the records of inspection and testing, together with certificates, have been completed and issued to the person ordering the work. Additionally, copies must also be provided to the building control body.

6.1.2 Building Regulations compliance certificate

Installation by a registered competent person

Within 30 days of the work being completed, the installer or installer's registration body must:

(a) give a copy of the Building Regulations compliance certificate to the occupier; and

(b) give the certificate, or a copy of the information on the certificate, to the building control body.

Certification by a registered third party

An installer who is not registered on a Competent Person Scheme may, before starting the work, appoint a registered third party certifier to inspect and test the work they carry out.

Within 5 days of completing the work, such an installer must notify the registered third party certifier. Subject to the work being satisfactory, the certifier should then give the completed IET inspection and testing reports (condition report) and certificate to the person ordering the work and notify their registration body.

Within 30 days of a satisfactory condition report being issued, the registration body of the third party certifier must:

(a) give a copy of the Building Regulations compliance certificate to the occupier; and

(b) give the certificate, or a copy of the information on the certificate, to the building control body.

Certification by a building control body

An installer who is not registered on a Competent Person Scheme and has not appointed a registered third party certifier, must notify a building control body before starting the work.

The building control body will decide the inspection and testing required by a third party to ensure that the work is safe, based on the nature of the work and the competence of the installer.

The building control body may decide to carry out any necessary inspection and testing itself, or it may contract a specialist to carry out some or all of the work. An installer who is competent to carry out inspection and testing should give the appropriate BS 7671 certificate to the building control body, who will then take the certificate and the installer's qualifications into account when deciding what further action, if any, it needs to take. Building control bodies may ask installers for evidence of their qualifications.

Once the building control body has decided that, as far as can be ascertained, the work meets all Building Regulations requirements, it will issue to the occupier a Building Regulations completion certificate (if a local authority) or a final certificate (if an approved inspector).

6.1.3   BS 7671 forms

Model forms are available on the Institution of Engineering and Technology website: www.theiet.org

When carrying out initial verification, inspection and testing on an installation, the short form Electrical Installation Certificate (Form 1) may be used together with Schedule of Inspections (Form 3) and a Schedule of Test Results (Form 4).

If other persons, including other companies, carry out the design and installation, the standard form Electrical Installation Certificate (Form 2) together with the Schedule of

Inspections and Schedule of Test Results is to be used. The information and signatures of skilled persons (electrically) for the design and construction should be completed prior to issuing the certificates.

BS 7671 requires:

(a) The Electrical Installation Certificate should be made out and signed or otherwise authenticated by an electrically skilled person or persons in respect of the design, construction, inspection and testing of the work.

(b) An electrically skilled person will have sound knowledge and experience relevant to the nature of the work undertaken, be fully versed in inspection and testing procedures and employ adequate testing equipment.

(c) Electrical Installation Certificates will indicate the responsibility for design, construction, inspection and testing, whether in relation to new work or to further work on an existing installation. Where the design, construction, inspection and testing are the responsibility of one person a certificate with a single-signature declaration in Form 1 may replace the multiple signatures section of the model form (Form 2).

(d) A Schedule of Inspections and a Schedule of Test Results must be issued with the Electrical Installation Certificate.

(e) When making out and signing a form on behalf of a company or other business entity, individuals must state for whom they are acting.

(f) Additional forms or pages may be required to permit explanation. Additional Schedules of Inspection and Schedules of Test Results may be needed for large or more complex installations.

The Electrical Installation Certificates reproduced here have been completed for typical installations.

▼ Form 1   Short form Electrical Installation Certificate

ELECTRICAL INSTALLATION CERTIFICATE   Form 1 No. ....*123*.../1
(REQUIREMENTS FOR ELECTRICAL INSTALLATIONS - BS 7671 [IET WIRING REGULATIONS])

DETAILS OF THE CLIENT
............*House Builder Ltd, 1 City Way, LONDON*..
                                                            Postcode: *EC3 3B*

INSTALLATION ADDRESS
............*Plot 25, New Road*
............*LONDON*
                                                            Postcode: *EC4 1A*

DESCRIPTION AND EXTENT OF THE INSTALLATION

Description of installation: *House*	New installation ☑
Extent of installation covered by this Certificate: *New installation including smoke and intruder alarms*	Addition to an existing installation ☐
	Alteration to an existing installation ☐
(Use continuation sheet if necessary) see continuation sheet No:	

FOR DESIGN, CONSTRUCTION, INSPECTION & TESTING
I/ being the person responsible for the design, construction, inspection & testing of the electrical installation (as indicated by my signature below), particulars of which are described above, having exercised reasonable skill and care when carrying out the design, construction, inspection & testing hereby CERTIFY that the design work for which I have been responsible is to the best of my/our knowledge and belief in accordance with BS 7671:2008, amended to *2015* (date) except for the departures, if any, detailed as follows:

Details of departures from BS 7671 (Regulations 120.3 and 133.5):
None

Details of permitted exceptions (Regulation 411.3.3). ^{Where applicable, a suitable risk assessment(s) must be attached to this Certificate.}
None
                                                    Risk assessment attached ☐

The extent of liability of the signatory is limited to the work described above as the subject of this Certificate.

Signature: ....*A. Smith*....   Date: ..*1/8/2015*.. Name (IN BLOCK LETTERS): ...*A. SMITH*...

PARTICULARS OF SIGNATORY
Position: *Electrician*   Company: ...*All Electric Ltd*...
Address: *27, Central Road*
            New Town.... Postcode: *AT4 5BA*.. Tel No: ..*04371/2166111*

NEXT INSPECTION
I recommend that this installation is further inspected and tested after an interval of not more than ...*5*...... years/~~months~~.

SUPPLY CHARACTERISTICS AND EARTHING ARRANGEMENTS

Earthing arrangements	Number and Type of Live Conductors		Nature of Supply Parameters	Supply Protective Device
TN-C ☐	a.c. ☑	d.c. ☐	Nominal voltage, U / U₀$^{(1)}$ ....230 V	BS (EN) *88*
TN-S ☐	1-phase, 2-wire ☑	2-wire ☐	Nominal frequency, f$^{(1)}$ ....50 Hz	Type *cartridge fuse*
TN-C-S ☑	2 phase, 3-wire ☐	3-wire ☐	Prospective fault current, I$_{pf}$$^{(2)}$ ...16 kA	Rated current 100 A
TT ☐	3 phase, 3-wire ☐	Other ☐	External loop impedance, Ze$^{(2)}$ 0.35 Ω	
IT ☐	3 phase, 4-wire ☐		*(Note: (1) by enquiry*	
	Confirmation of supply polarity ☑		*(2) by enquiry or by measurement)*	
Other sources of supply (as detailed on attached schedule) ☐ *None*				

                                                            Page 1 of .*5*..

6

▼ **Form 1** *continued*

PARTICULARS OF INSTALLATION REFERRED TO IN THE CERTIFICATE		
Means of Earthing	**Maximum Demand**	
Distributor's facility ☑	Maximum demand (load) .... 60 kVA / Amps Delete as appropriate	
Installation earth electrode ☐	**Details of Installation Earth Electrode** *(where applicable)*	
	Type (e.g. rod(s), tape etc).......... N/A	
	Location .... N/A	
	Electrode resistance to Earth .....N/A.... Ω	

Main Protective Conductors				
Earthing conductor	Material ....Copper.... csa .....16... mm²		Connection / continuity verified	☑
Main protective bonding conductors (to extraneous-conductive-parts)	Material ....Copper.... csa .....10... mm²		Connection / continuity verified	☑
To water installation pipes ☑	To gas installation pipes ☑	To oil installation pipes N/A	To structural steel N/A	
To lightning protection N/A	To other ☐ Specify .....N/A			

Main Switch / Switch-Fuse / Circuit-Breaker / RCD		
Location .....Garage.....	Current rating 100.A	**If RCD main switch**
	Fuse / device rating or setting N/A.A	Rated residual operating current (IΔn) ....mA
BS(EN) ...60947-3...	Voltage rating 230.V	Rated time delay ....ms
No of poles .....2.....		Measured operating time(at IΔn) ....ms

COMMENTS ON EXISTING INSTALLATION (in the case of an addition or alteration see Section 633):

New installation

SCHEDULES
The attached Schedules are part of this document and this Certificate is valid only when they are attached to it.
.....1..... Schedules of Inspections and ....1..... Schedules of Test Results are attached.
(Enter quantities of schedules attached).

ELECTRICAL INSTALLATION CERTIFICATE
GUIDANCE FOR RECIPIENTS (to be appended to the Certificate)

This safety Certificate has been issued to confirm that the electrical installation work to which it relates has been designed, constructed, inspected and tested in accordance with British Standard 7671 (the IET Wiring Regulations).

You should have received an "original" Certificate and the contractor should have retained a duplicate. If you were the person ordering the work, but not the owner of the installation, you should pass this Certificate, or a full copy of it including the schedules, immediately to the owner.

The "original" Certificate should be retained in a safe place and be shown to any person inspecting or undertaking further work on the electrical installation in the future. If you later vacate the property, this Certificate will demonstrate to the new owner that the electrical installation complied with the requirements of British Standard 7671 at the time the Certificate was issued. The Construction (Design and Management) Regulations require that, for a project covered by those Regulations, a copy of this Certificate, together with schedules, is included in the project health and safety documentation.

For safety reasons, the electrical installation will need to be inspected at appropriate intervals by a skilled person or persons, competent in such work. The maximum time interval recommended before the next inspection is stated on Page 1 under "NEXT INSPECTION".

This Certificate is intended to be issued only for a new electrical installation or for new work associated with an addition or alteration to an existing installation. It should not have been issued for the inspection of an existing electrical installation. An "Electrical Installation Condition Report" should be issued for such an inspection.

This Certificate is only valid if accompanied by the Schedule of Inspections and the Schedule(s) of Test Results.

Page 2 of .5...

▼ **Form 2**   Standard form Electrical Installation Certificate

ELECTRICAL INSTALLATION CERTIFICATE
(REQUIREMENTS FOR ELECTRICAL INSTALLATIONS - BS 7671 [IET WIRING REGULATIONS])   Form 2 No. .... *124* /2

DETAILS OF THE CLIENT
A Developer Ltd, Main Street, LONDON
Postcode: *SW1 2BE*

INSTALLATION ADDRESS
55, High Street
Town
County Postcode: *AW2 4CE*

DESCRIPTION AND EXTENT OF THE INSTALLATION

Description of installation: *Shop and house*	New installation	☑
Extent of installation covered by this Certificate: *Complete installation*	Addition to an existing installation	☐
	Alteration to an existing installation	☐
(Use continuation sheet if necessary) see continuation sheet No:		

FOR DESIGN

I/We being the person(s) responsible for the design of the electrical installation (as indicated by my/our signatures below), particulars of which are described above, having exercised reasonable skill and care when carrying out the design hereby CERTIFY that the design work for which I/we have been responsible is to the best of my/our knowledge and belief in accordance with BS 7671:2008, amended to *2015* (date) except for the departures, if any, detailed as follows:

Details of departures from BS 7671 (Regulations 120.3 and 133.5):   *None*

Details of permitted exceptions (Regulation 411.3.3). Where applicable, a suitable risk assessment(s) must be attached to this Certificate.

None

Risk assessment attached ☐

The extent of liability of the signatory or signatories is limited to the work described above as the subject of this Certificate.

For the DESIGN of the installation: **(Where there is mutual responsibility for the design)

Signature: *B Brown*   Date: *1/8/2015*   Name (IN BLOCK LETTERS): *B. BROWN*   Designer No 1

Signature: ....   Date: ....   Name (IN BLOCK LETTERS): ....   Designer No 2**

FOR CONSTRUCTION

I being the person responsible for the construction of the electrical installation (as indicated by my signature below), particulars of which are described above, having exercised reasonable skill and care when carrying out the construction hereby CERTIFY that the construction work for which I have been responsible is to the best of my knowledge and belief in accordance with BS 7671:2008, amended to *2015* (date) except for the departures, if any, detailed as follows:

Details of departures from BS 7671 (Regulations 120.3 and 133.5):   *None*

The extent of liability of the signatory is limited to the work described above as the subject of this Certificate.

For CONSTRUCTION of the installation:

Signature: *W White*   Date: *7/12/2015*   Name (IN BLOCK LETTERS): *W. WHITE*   Constructor

FOR INSPECTION & TESTING

I being the person responsible for the inspection & testing of the electrical installation (as indicated by my signature below), particulars of which are described above, having exercised reasonable skill and care when carrying out the inspection & testing hereby CERTIFY that the work for which I have been responsible is to the best of my knowledge and belief in accordance with BS 7671:2008, amended to *2015* (date) except for the departures, if any, detailed as follows:

Details of departures from BS 7671 (Regulations 120.3 and 133.5):   *None*

The extent of liability of the signatory is limited to the work described above as the subject of this Certificate.

For INSPECTION AND TESTING of the installation:

Signature: *S Jones*   Date: *9/12/2015*   Name (IN BLOCK LETTERS): *S. JONES*   Inspector

NEXT INSPECTION

I/We the designer(s), recommend that this installation is further inspected and tested after an interval of not more than ....*5*..... years/~~months~~.

Page 1 of .*5*..

▼ Form 2 *continued*

Form 2 No. ....**124**... /2

PARTICULARS OF SIGNATORIES TO THE ELECTRICAL INSTALLATION CERTIFICATE	
Designer (No 1) Name: **B Brown** Address: **City Road** **Old Town**	Company: **Design Co Ltd** Postcode: **PB4 7G** Tel No: **01234 5678**
Designer (No 2) (if applicable) Name: **----** Address:	Company: **----** Postcode: Tel No:
Constructor Name: **W White** Address: **187, Long Lane** **Town**	Company: **County Electrics** Postcode: **PB6 8HJ** Tel No: **01234 4765**
Inspector Name: **S Jones** Address: **187, Long lane** **Town**	Company: **County Electrics** Postcode: **PB6 8HJ** Tel No: **01234 4868**

SUPPLY CHARACTERISTICS AND EARTHING ARRANGEMENTS

Earthing arrangements	Number and Type of Live Conductors		Nature of Supply Parameters	Supply Protective Device
TN-C ☐	a.c. ☑	d.c. ☐	Nominal voltage, U / U_0 [(1)] **400/230** V	BS (EN) ....**88**....
TN-S ☐	1-phase, 2-wire ☐	2-wire ☐	Nominal frequency, f [(1)] ....**50** Hz	Type ....**fuse**....
TN-C-S ☑	2 phase, 3-wire ☐	3-wire ☐	Prospective fault current, I_{pf} [(2)] .....**18** kA	Rated current ....**100**.... A
TT ☐	3 phase, 3-wire ☐	Other ☐	External loop impedance, Z_e [(2)] ...**0.2** Ω	
IT ☐	3 phase, 4-wire ☑		*(Note: (1) by enquiry*	
	Confirmation of supply polarity ☐		*(2) by enquiry or by measurement)*	
Other sources of supply (as detailed on attached schedule) ☐				

PARTICULARS OF INSTALLATION REFERRED TO IN THE CERTIFICATE

Means of Earthing	**Maximum Demand**
Distributor's facility ☑	Maximum demand (load) ....**40** kVA / Amps *Delete as appropriate*
Installation earth electrode ☐	**Details of Installation Earth Electrode** *(where applicable)* Type (e.g. rod(s), tape etc) .... Location .... Electrode resistance to Earth .... Ω

Main Protective Conductors

Earthing conductor	Material ......**Copper**...... csa ......**16**... mm²	Connection / continuity verified ☑
Main protective bonding conductors (to extraneous-conductive-parts)	Material ......**Copper**...... csa ......**10**... mm²	Connection / continuity verified ☑

To water installation pipes ☑	To gas installation pipes ☑	To oil installation pipes **N/A**	To structural steel **N/A**
To lightning protection **N/A**	To other ☐ Specify ....**N/A**....		

Main Switch / Switch-Fuse / Circuit-Breaker / RCD

Location **Cupboard near** **to entrance** BS(EN) ......**60947-3**...... No of poles **3**	Current rating ....**125** A Fuse / device rating or setting ==.=. A Voltage rating ....**400** V	**If RCD main switch** Rated residual operating current ($I_{Δn}$) ....mA Rated time delay ....ms Measured operating time (at $I_{Δn}$) ....ms

COMMENTS ON EXISTING INSTALLATION (in the case of an addition or alteration see Section 633):

Not applicable, new installation

SCHEDULES

The attached Schedules are part of this document and this Certificate is valid only when they are attached to it.
......**1**... Schedules of Inspections and ....**1**.... Schedules of Test Results are attached.
(Enter quantities of schedules attached)

Page 2 of ..**5**...

ELECTRICAL INSTALLATION CERTIFICATE NOTES:

1   The Electrical Installation Certificate is to be used only for the initial certification if a new installation or for an addition or alteration to an existing installation where new circuits have been introduced.
    It is not used for a Periodic Inspection, for which an Electrical Installation Condition Report form should be used. For an addition or alteration, which does not extend to the introduction of new circuits, a Minor Electrical Installation Works Certificate may be used.
    The 'original' Certificate is to be given to the person ordering the work (Regulation 632.1). A duplicate should be retained by the contractor.

2   This Certificate is only valid if accompanied by the Schedule of Inspections and the Schedule(s) of Test Results.

3   The signatures appended are those of the persons authorised by the companies executing the work of design, construction, inspection and testing retrospectively. A signatory authorised to certify more than one category of work should sign in each of the appropriate places.

4   The time interval recommended before the first periodic inspection must be inserted.
    The proposed date for the next inspection should take into consideration the frequency and quality of maintenance that the installation can reasonably be expected to receive during its intended life, and the period should be agreed between the designer, installer and other relevant parties.

5   The page numbers for each of the Schedules of Test Results should be indicated, together with the total number of sheets involved.

6   The maximum prospective value of fault current (I_{pf}) recorded should be the greater of either the prospective value of short-circuit current or the prospective value of earth fault current.

ELECTRICAL INSTALLATION CERTIFICATES

Guidance for recipients of Form 1 or Form 2 (to be appended to the Certificate):

This safety Certificate has been issued to confirm that the electrical installation work to which it relates has been designed, constructed, inspected and tested in accordance with British Standard 7671 (the IET Wiring Regulations).

You should have received an 'original' Certificate and the contractor should have retained a duplicate. If you were the person ordering the work, but not the owner of the installation, you should pass this Certificate, or a full copy of it including the schedules, immediately to the owner.

The 'original' Certificate should be retained in a safe place and be shown to any person inspecting or undertaking further work on the electrical installation in the future. If you later vacate the property, this Certificate will demonstrate to the new owner that the electrical installation complied with the requirements of British

Standard 7671 at the time the Certificate was issued. The Construction (Design and Management) Regulations require that, for a project covered by those Regulations, a copy of this Certificate, together with schedules, is included in the project health and safety documentation.

For safety reasons, the electrical installation will need to be inspected at appropriate intervals by an electrically skilled person or persons, competent in such work. The maximum time interval recommended before the next inspection is stated on Page 1 under 'Next Inspection'.

This Certificate is intended to be issued only for a new electrical installation or for new work associated with an addition or alteration to an existing installation. It should not have been issued for the inspection of an existing electrical installation. An 'Electrical Installation Condition Report' should be issued for such an inspection.

This Certificate is only valid if accompanied by the Schedule of Inspections and the Schedule(s) of Test Results.

6.2   Inspection

6.2.1   General

Every installation must be inspected during erection as necessary and on completion and before being put into service to provide a visual check that the installation including the installed equipment complies with the requirements of BS 7671.

The inspection will check that the equipment is:

   (a) made in compliance with the appropriate British Standards or European Standards;
   (b) selected and installed in accordance with BS 7671 (including consideration of external influences such as the presence of moisture); and
   (c) not visibly damaged or defective so as to be unsafe.

Inspection must precede testing and is normally to be done with the part of the installation that is under inspection disconnected from the supply.

Defects or omissions revealed during inspection of the installation work covered by the Schedule of Inspection (Form 3) must be made good before the Electrical Installation Certificate is issued. If the inspection reveals no departures it must be signed in preparation for giving to the person ordering the work as part of the set of forms.

Examples of items requiring inspection during initial verification

All items inspected in order to confirm, as appropriate, compliance with the relevant clauses in BS 7671.

The list of items is not exhaustive. Numbers in brackets are Regulation references.

ELECTRICAL INTAKE EQUIPMENT

- Service cable
- Service head
- Distributor's earthing arrangement
- Meter tails - Distributor/Consumer
- Metering equipment
- Isolator

PARALLEL OR SWITCHED ALTERNATIVE SOURCES OF SUPPLY

- Presence of adequate arrangements where generator to operate as a switched alternative (551.6)
  1. Dedicated earthing arrangement independent of that of the public supply (551.4.3.2.1)

- Presence of adequate arrangements where generator to operate in parallel with the public supply system (551.7)
  1. Correct connection of generator in parallel (551.7.2)
  2. Compatibility of characteristics of means of generation (551.7.3)
  3. Means to provide automatic disconnection of generator in the event of loss of public supply system or voltage or frequency deviation beyond declared values (551.7.4)
  4. Means to prevent connection of generator in the event of loss of public supply system or voltage or frequency deviation beyond declared values (551.7.5)
  5. Means to isolate generator from the public supply system (551.7.6)

AUTOMATIC DISCONNECTION OF SUPPLY

- Protective earthing/protective bonding srrangements (411.3; Chap 54)
- resence and adequacy of
  1. Distributor's earthing arrangement (542.1.2.1; 542.1.2.2), or installation earth electrode arrangement (542.1.2.3)
  2. Earthing conductor and connections (Section 526; 542.3; 542.3.2; 543.1.1)
  3. Main protective bonding conductors and connections (Section 526; 544.1; 544.1.2)
  4. Earthing/bonding labels at all appropriate locations (514.13)
- Accessibility of
  1. Earthing conductor connections
  2. All protective bonding connections (543.3.2)
- FELV - requirements satisfied (411.7; 411.7.1)

OTHER METHODS OF PROTECTION (Where any methods listed below are employed details should be provided on separate pages)

BASIC AND FAULT PROTECTION where used, confirmation that the requirements are satisfied:

- SELV (Section 414)
- PELV (Section 414)
- Double insulation (Section 412)
- Reinforced insulation (Section 412)

BASIC PROTECTION:

- Insulation of live parts (416.1)
- Barriers or enclosures (416.2; 416.2.1)
- Obstacles (Section 417; 417.2.1; 417.2.2)
- Placing out of reach (Section 417; 417.3)

FAULT PROTECTION:

- Non-conducting location (418.1)
- Earth-free local equipotential bonding (418.2)
- Electrical separation (Section 413; 418.3)

ADDITIONAL PROTECTION:

- RCDs not exceeding 30 mA as specified (411.3.3; 415.1)
- Supplementary bonding (Section 415; 415.2)

SPECIFIC INSPECTION EXAMPLES as appropriate to the installation

DISTRIBUTION EQUIPMENT

- Security of fixing (134.1.1)
- Insulation of live parts not damaged during erection (416.1)
- Adequacy/security of barriers (416.2)
- Suitability of enclosures for IP and fire ratings (416.2; 421.1.6; 421.1.201; 526.5)
- Enclosures not damaged during installation (134.1.1)
- Presence and effectiveness of obstacles (417.2)
- Presence of main switch(es), linked where required (537.1.3; .4; .5; .6)
- Operation of main switch(es) (functional check) (612.13)
- Manual operation of circuit-breakers and RCDs to prove functionality (612.13.2)
- Confirmation that integral test button/switch causes RCD(s) to trip when operated (functional check) (612.13.1)
- RCD(s) provided for fault protection, where specified (411.4.9; 411.5.2; 531.2)
- RCD(s) provided for additional protection, where specified (411.3.3; 415.1)
- Confirmation overvoltage protection (SPDs) provided where specified (534.2.1)
- Confirmation of indication that SPD is functional (534.2.8)
- Presence of RCD quarterly test notice at or near the origin (514.12.2)

- Presence of diagrams, charts or schedules at or near each distribution board, where required (514.9.1)
- Presence of non-standard (mixed) cable colour warning notice at or near the appropriate distribution board, where required (514.14)
- Presence of alternative supply warning notice at or near (514.15)
  1. The origin
  2. The meter position, if remote from origin
  3. The distribution board to which the alternative/additional sources are connected
  4. All points of isolation of ALL sources of supply
- Presence of next inspection recommendation label (514.12.1)
- Presence of other required labelling (Section 514)
- Selection of protective device(s) and base(s); correct type and rating (411.3.2; 411.4, .5, .6; Sections 432, 433)
- Single-pole protective devices in line conductors only (132.14.1, 530.3.2)
- Protection against mechanical damage where cables enter equipment (522.8.1; 522.8.11)
- Protection against electromagnetic effects where cables enter ferromagnetic enclosures (521.5.1)
- Confirmation that ALL conductor connections, including connections to busbars, are correctly located in terminals and are tight and secure (526.1)

CIRCUITS

- Identification of conductors (514.3.1)
- Cables correctly supported throughout (522.8.5)
- Examination of cables for signs of mechanical damage during installation (522.6.1; 522.8.1)
- Examination of insulation of live parts, not damaged during erection (522.6.1; 522.8.1)
- Non-sheathed cables protected by enclosure in conduit, ducting or trunking (521.10.1)
- Suitability of containment systems (including flexible conduit) (Section 522)
- Correct temperature rating of cable insulation (522.1.1; Table 52.1)
- Adequacy of cables for current-carrying capacity with regard for the type and nature of installation (Section 523)
- Adequacy of protective devices: type and fault current rating for fault protection (434.5)
- Presence and adequacy of circuit protective conductors (411.3.1; 543.1)
- Coordination between conductors and overload protective devices (433.1; 533.2.1)
- Wiring systems and cable installation methods/practices with regard to the type and nature of installation and external influences (Section 522)
- Cables concealed under floors, above ceilings, in walls/partitions, adequately protected against damage (522.6.201, .202, .204)
- Provision of additional protection by RCDs having rated residual operating current ($I_{\Delta n}$) not exceeding 30 mA

1. For circuits used to supply mobile equipment not exceeding 32 A rating for use outdoors (411.3.3)
2. For all socket-outlets of rating 20 A or less, unless exempt (411.3.3)
3. For cables concealed in walls at a depth of less than 50 mm (522.6.202, .203)
4. For cables concealed in walls/partitions containing metal parts regardless of depth (522.6.202, .203)

- Provision of fire barriers, sealing arrangements so as to minimize the spread of fire (Section 527)
- Band II cables segregated/separated from Band I cables (528.1)
- Cables segregated/separated from non-electrical services (528.3)
- Termination of cables at enclosures (Section 526)
   1. Connections under no undue strain (526.6)
   2. No basic insulation of a conductor visible outside enclosure (526.8)
   3. Connections of live conductors adequately enclosed (526.5)
   4. Adequately connected at point of entry to enclosure (glands, bushes etc.) (522.8.5)
- Suitability of circuit accessories for external influences (512.2)
- Circuit accessories not damaged during erection (134.1.1)
- Single-pole devices for switching or protection in line conductors only (132.14.1, 530.3.2)
- Adequacy of connections, including cpc's, within accessories and at fixed and stationary equipment (Section 526)

ISOLATION AND SWITCHING

- Isolators (537.2)
   1. Presence and location of appropriate devices (537.2.2)
   2. Capable of being secured in the OFF position (537.2.1.2)
   3. Correct operation verified (functional check) (612.13.2)
   4. The installation, circuit or part thereof that will be isolated clearly identified by location and/or durable marking (537.2.2.6)
   5. Warning notice posted in situation where live parts cannot be isolated by the operation of a single device (514.11.1; 537.2.1.3)
- Switching off for mechanical maintenance (537.3)
   1. Presence of appropriate devices (537.3.1.1)
   2. Acceptable location – state if local or remote from equipment in question (537.3.2.4)
   3. Capable of being secured in the OFF position (537.3.2.3)
   4. Correct operation verified (functional check) (612.13.2)
   5. The circuit or part thereof to be disconnected clearly identified by location and/or durable marking (537.3.2.4)
- Emergency switching/stopping (537.4)
   1. Presence of appropriate devices (537.4.1.1)
   2. Readily accessible for operation where danger might occur (537.4.2.5)
   3. Correct operation verified (functional check) (537.4.2.6)
   4. The installation, circuit or part thereof to be disconnected clearly identified by location and/or durable marking (537.4.2.7)

- Functional switching (537.5)
    1. Presence of appropriate devices (537.5.1.1)
    2. Correct operation verified (functional check) (537.5.1.3; 537.5.2.2)

CURRENT–USING EQUIPMENT (PERMANENTLY CONNECTED)

- Suitability of equipment in terms of IP and fire ratings (416.2)
- Enclosure not damaged/deteriorated during installation so as to impair safety (134.1.1)
- Suitability for the environment and external influences (512.2)
- Security of fixing (134.1.1)
- Cable entry holes in ceilings above luminaires, sized or sealed so as to restrict the spread of fire
- Provision of undervoltage protection, where specified (Section 445)
- Provision of overload protection, where specified (Section 433; 552.1)
- Recessed luminaires (downlighters)
    1. Correct type of lamps fitted
    2. Installed to minimize build-up of heat (421.1.2; 559.4.1)
- Adequacy of working space/accessibility to equipment (132.12; 513.1)

PART 7 SPECIAL INSTALLATIONS OR LOCATIONS

- Particular requirements for special locations are fulfilled.

A sample Form 3 Schedule of Inspections shown below has been completed for a typical domestic dwelling. Blank forms are available on the IET website.

▼ **Form 3**   Schedule of Inspections (to accompany either the short or standard form Electrical Installation Certificate)

SCHEDULE OF INSPECTIONS (for new installation work only) for DOMESTIC AND SIMILAR PREMISES WITH UP TO 100 A SUPPLY   Form 3 No. .... *123* ../3

NOTE 1: This form is suitable for many types of smaller installation, not exclusively domestic.

All items inspected in order to confirm, as appropriate, compliance with the relevant clauses in BS 7671. The list of items and associated examples where given are not exhaustive.

NOTE 2: Insert ✓ to indicate an inspection has been carried out and the result is satisfactory,
or N/A to indicate that the inspection is not applicable to a particular item.

ITEM NO	DESCRIPTION	Outcome See Note 2
1.0	**DISTRIBUTOR'S / SUPPLY INTAKE EQUIPMENT**	
1.1	Condition of service cable	✓
1.2	Condition of service head	✓
1.3	Condition of distributor's earthing arrangement	✓
1.4	Condition of meter tails - Distributor/Consumer	✓
1.5	Condition of metering equipment	✓
1.6	Condition of isolator (where present)	N/A
2.0	**PARALLEL OR SWITCHED ALTERNATIVE SOURCES OF SUPPLY**	
2.1	Adequate arrangements where a generating set operates as a switched alternative to the public supply (551.6)	N/A
2.2	Adequate arrangements where a generating set operates in parallel with the public supply (551.7)	N/A
3.0	**AUTOMATIC DISCONNECTION OF SUPPLY**	
3.1	**Presence and adequacy of earthing and protective bonding arrangements:**	
	• Installation earth electrode (where applicable) (542.1.2.3)	N/A
	• Earthing conductor and connections, including accessibility (542.3; 543.3.2)	✓
	• Main protective bonding conductors and connections, including accessibility (411.3.1.2; 543.3.2)	✓
	• Provision of safety electrical earthing / bonding labels at all appropriate locations (514.13)	✓
	• RCD(s) provided for fault protection (411.4.9; 411.5.3)	N/A
4.0	**BASIC PROTECTION**	
4.1	**Presence and adequacy of measures to provide basic protection (prevention of contact with live parts) within the installation:**	
	• Insulation of live parts e.g. conductors completely covered with durable insulating material (416.1)	✓
	• Barriers or enclosures e.g. correct IP rating (416.2)	✓
5.0	**ADDITIONAL PROTECTION**	
5.1	**Presence and effectiveness of additional protection methods:**	
	•RCD(s) not exceeding 30 mA operating current (415.1; Part 7), see Item 8.14 of this schedule	✓
	•Supplementary bonding (415.2; Part 7)	✓
6.0	**OTHER METHODS OF PROTECTION**	
6.1	**Presence and effectiveness of methods which give both basic and fault protection:**	
	• SELV system, including the source and associated circuits (Section 414)	N/A
	• PELV system, including the source and associated circuits (Section 414)	N/A
	• Double or reinforced insulation i.e. Class II or equivalent equipment and associated circuits (Section 412)	N/A
	• Electrical separation for one item of equipment e.g. shaver supply unit (Section 413)	N/A
7.0	**CONSUMER UNIT(S) / DISTRIBUTION BOARD(S):**	
7.1	Adequacy of access and working space for items of electrical equipment including switchgear (132.12)	✓
7.2	Presence of linked main switch(s) (537.1.4; 537.1.5; 537.1.6)	✓
7.3	Isolators, for every circuit or group of circuits and all items of equipment (537.2)	✓
7.4	Suitability of enclosure(s) for IP and fire ratings (416.2; 421.1.6; 421.1.201)	✓

Page *3*.. of ..*5*..

Form 3 No. ....**123**...../3

ITEM NO	DESCRIPTION	Outcome See Note 2
	CONSUMER UNIT(S) / DISTRIBUTION BOARD(S) continued	
7.5	Protection against mechanical damage where cables enter equipment (522.8.1; 522.8.11)	✓
7.6	Confirmation that ALL conductor connections are correctly located in terminals and are tight and secure (526.1)	✓
7.7	Avoidance of heating effects where cables enter ferromagnetic enclosures e.g. steel (521.5)	✓
7.8	Selection of correct type and ratings of circuit protective devices for overcurrent and fault protection (411.3.2; 411.4, .5, .6; Sections 432, 433)	✓
7.9	**Presence of appropriate circuit charts, warning and other notices:**	
	• Provision of circuit charts/schedules or equivalent forms of information (514.9)	✓
	• Warning notice of method of isolation where live parts not capable of being isolated by a single device (514.11)	✓
	• Periodic inspection and testing notice (514.12.1)	✓
	• RCD quarterly test notice; where required (514.12.2)	✓
	• Warning notice of non-standard (mixed) colours of conductors present (514.14)	N/A
7.10	Presence of labels to indicate the purpose of switchgear and protective devices (514.1.1; 514.8)	✓
8.0	**CIRCUITS**	
8.1	Adequacy of conductors for current-carrying capacity with regard to type and nature of the installation (Section 523)	✓
8.2	Cable installation methods suitable for the location(s) and external influences (Section 522)	✓
8.3	Segregation/separation of Band I (ELV) and Band II (LV) circuits, and electrical and non-electrical services (528)	✓
8.4	Cables correctly erected and supported throughout including escape routes, with protection against abrasion (Sections 521, 522)	✓
8.5	Provision of fire barriers, sealing arrangements where necessary (527.2)	
8.6	Non-sheathed cables enclosed throughout in conduit, ducting or trunking (521.10.1; 526.8)	✓
8.7	Cables concealed under floors, above ceilings or in walls/partitions, adequately protected against damage (522.6.201, .202, .204)	✓
8.8	Conductors correctly identified by colour, lettering or numbering (Section 514)	✓
8.9	Presence, adequacy and correct termination of protective conductors (411.3.1.1; 543.1)	✓
8.10	Cables and conductors correctly connected, enclosed and with no undue mechanical strain (Section 526)	✓
8.11	No basic insulation of a conductor visible outside enclosure (526.8)	✓
8.12	Single-pole devices for switching or protection in line conductors only (132.14.1; 530.3.2)	✓
8.13	Accessories not damaged, securely fixed, correctly connected, suitable for external influences (134.1.1; 512.2; Section 526)	✓
8.14	**Provision of additional protection by RCD not exceeding 30mA:**	
	• Socket-outlets rated at 20 A or less, unless exempt (411.3.3)	✓
	• Mobile equipment with a current rating not exceeding 32 A for use outdoors (411.3.3)	✓
	• Cables concealed in walls at a depth of less than 50 mm (522.6.202, .203)	✓
	• Cables concealed in walls / partitions containing metal parts regardless of depth (522.6.202; 522.6.203)	✓
8.15	**Presence of appropriate devices for isolation and switching correctly located including:**	
	• Means of switching off for mechanical maintenance (537.3)	✓
	• Emergency switches (537.4)	N/A
	• Functional switches, for control of parts of the installation and current-using equipment (537.5)	✓
	• Firefighter's switches (537.6)	N/A
9.0	**CURRENT-USING EQUIPMENT (PERMANENTLY CONNECTED)**	
9.1	Equipment not damaged, securely fixed and suitable for external influences (134.1.1; 416.2; 512.2)	✓
9.2	Provision of overload and/or undervoltage protection e.g. for rotating machines, if required (Sections 445, 552)	N/A
9.3	Installed to minimize the build-up of heat and restrict the spread of fire (421.1.4; 559.4.1)	✓
9.4	Adequacy of working space. Accessibility to equipment (132.12; 513.1)	✓
10.0	**LOCATION(S) CONTAINING A BATH OR SHOWER (SECTION 701)**	
10.1	30 mA RCD protection for all LV circuits, equipment suitable for the zones, supplementary bonding (where required) etc.	✓
11.0	**OTHER PART 7 SPECIAL INSTALLATIONS OR LOCATIONS**	
11.1	List all other special installations or locations present, if any. (Record separately the results of particular inspections applied)	✓

Inspected by:
Name (Capitals) ...A. SMITH.......... Signature ....*A Smith*........ Date 8/12/2015

Page ..4.. of ..5..

6.3   Testing

6.3.1   Safety and equipment

(a) Electrical testing involves a degree of risk. The tester is responsible not only for his or her own safety, but also for the safety of others.

(b) When design and construction have been carried out by others, a standard Form 2 signed by the electrically skilled persons responsible for the design and construction should be in the possession of the inspector before starting inspection and testing.

(c) Inspection always precedes testing.

(d) Dead tests always precede live tests.

(e) On each occasion before using test equipment, the tester must confirm that all leads, probes, accessories (including all devices such as crocodile clips used to attach conductors) and instruments are undamaged and are functional.

(f) Manufacturer's instructions must be read before using equipment and must thereafter be followed. It should be noted that some test instrument manufacturers advise that their instruments be used in conjunction with fused test leads and probes.

GS 38

Safety procedures are described in Health and Safety Executive Guidance Note GS 38 *Electrical test equipment for use by electricians.* Test instruments should comply with the BS EN 61010 series *Safety requirements for electrical equipment for measurement, control and laboratory use.*

Test instruments and in particular test leads including prods and clips should be in good order. Fused test leads (to reduce the risks associated with arcing under fault conditions) must be used when recommended by the manufacturer.

6.3.2   Schedule of Test Results

The Schedule of Test Results must be available during testing, must identify every circuit, including its related protective device(s), and must record the results of the appropriate tests and measurements.

A sample completed Schedule is shown opposite (Form 4).

▶ **Form 4**   Schedule of Test Results

GENERIC SCHEDULE OF TEST RESULTS

Form 4 No. ....123..... /4

DB reference no *Consumer unit*	
Location *Hall*	
Zs at DB (Ω) *0.4*	
Ipf at DB (kA) *0.6*	
Correct supply polarity confirmed ☑	
Phase sequence confirmed (where appropriate) ☐	

Details of circuits and/or installed equipment vulnerable to damage when testing *Electronic shower control*   *E.L.V. lighting*

Details of test instruments used (state serial and/or asset numbers)
- Continuity .... *AB 11*
- Insulation resistance .... *AB 22*
- Earth fault loop impedance .... *AB 33*
- RCD .... *AB 44*
- Earth electrode resistance .... *- - -*

Tested by:
Name (Capitals) ..... *A. SMITH*
Signature ..... *A. Smith* ..... Date *8/12/2015*

Circuit number	Circuit Description	Overcurrent device				Conductor details			Ring final circuit continuity (Ω)			Continuity (Ω) (R₁+R₂) or R₂		Insulation Resistance (MΩ)		Polarity	Zs (Ω)	RCD (ms)			Remarks (continue on a separate sheet if necessary)
		BS (EN)	type	rating (A)	breaking capacity (kA)	Reference Method	Live (mm²)	cpc (mm²)	r_1 (line)	r_n (neutral)	r_2 (cpc)	$(R_1 + R_2)$	R_2	Live - Live	Live - Earth	Insert ✓ or X		$I_{\Delta n}$	$5I_{\Delta n}$	Test button operation	
1	2	3	4	5	6	7	8	9	10	11	12	13	14	15	16	17	18	19	20	21	22
	RCD 1																				
1	Lights up	60898 B		10	6	100	1.5	1.0	-	-	-	2.4	-	100	100	✓	2.8	35	18	✓	
2	Sockets down	60898 B		32	6	102	2.5	1.5	0.74	0.74	1.2	0.5	-	90	90	✓	0.9				Dimmer
3	Cooker	60898 B		32	6	102	6.0	2.5	-	-	-	0.35	-	100	100	✓	0.9				
	RCD 2																				
4	Lights down	60898 B		10	6	100	1.5	1.0	-	-	-	2.4	-	100	100	✓	3.2	39	17	✓	
5	Sockets up	60898 B		32	6	102	2.5	1.5	0.59	0.59	0.96	0.4	-	90	90	✓	0.8				
6	Shower	60898 B		45	6	102	10	4.0	-	-	-	0.35	-	100	100	✓	0.6				Electronic

Page *5* of *5*

GENERIC SCHEDULE OF TEST RESULTS

NOTES

The following notes relate to the column number in the form (Form 4).

1   Circuit number – for three-phase installations it is preferred to use the designation LI, L2, L3 – so (for example) for the 5th circuit, the designation 5L1, 5L2 and 5L3.

2   Circuit description – can be brief, e.g. fluorescent lighting; see completed sheet.

3   BS (EN) – enter the Standard of manufacture of the circuit protective device, e.g. (BS EN) 60898.

4   Type – where relevant for circuit-breakers enter the sensitivity type, e.g. C.

5   Rating – enter the protective device current rating.

6   Breaking capacity – enter the protective device breaking capacity, often 'printed' on circuit-breakers, e.g. 6.

7   Reference Method – enter the cable's installed reference method, by using Table 4A2 of BS 7671.

8   Conductor size – enter live conductor size in mm^2.

9   Conductor size – enter circuit protective conductor size in mm^2.

10   Ring line – line open resistance continuity in ohms, see 6.3.5.

11   Ring neutral – neutral open resistance continuity in ohms, see 6.3.5.

12   Ring cpc – cpc open resistance continuity in ohms, see 6.3.5.

13   Ring ($R_1 + R_2$) – enter the value recorded whilst carrying out step 3 of the ring continuity test, see 6.3.5. Note that where senseless results are recorded, due to parallel return paths, and it has been established and the inspector has verified continuity, a value is not necessary in this cell, and the cell may be ticked.

14   Continuity R_2 – add the value of the cpc continuity reading. If using test method 2, the 'wandering lead' method, then enter the maximum value of the various readings that were measured on the circuit. Note that where senseless results are recorded, due to parallel return paths, and it has been established and the inspector has verified continuity, a value is not necessary in this cell, and the cell may be ticked.

15   Insulation, live–live – enter the minimum value recorded during testing the circuit for each of the various configurations, see 6.3.6.

16   Insulation, live–Earth – enter the minimum value recorded during testing the circuit for each of the various configurations, see 6.3.6.

17   Polarity – tick this cell when the polarity for the circuit has been confirmed, see 6.3.7.

18   Z_s – enter the circuit earth fault loop impedance by whatever method you have selected to determine it by, see 6.3.9.

19, 20 and 21   Enter the results from the tests carried out on any RCDs fitted to the circuit, see 6.4.

22   Remarks – this cell is provided to note anything relevant to the circuit and testing.

6.3.3 Sequence of tests

Tests should be carried out in the following sequence:

Before the supply is connected:

- **(a)** continuity of protective conductors, including main and supplementary bonding;
- **(b)** continuity of ring final circuit conductors, including protective conductors;
- **(c)** insulation resistance;
- **(d)** polarity (by continuity methods); and
- **(e)** earth electrode resistance, using an earth electrode resistance tester (see also (g)).

With the supply connected and energized:

- **(f)** re-check polarity, using an approved voltage indicator;
- **(g)** earth electrode resistance, using a loop impedance tester;
- **(h)** earth fault loop impedance;
- **(i)** prospective fault current measurement, if not determined by enquiry of the distributor; and
- **(j)** functional testing, including RCDs and switchgear.

Results obtained during tests should be recorded in the Schedule of Test Results.

6.3.4 Continuity of protective and bonding conductors

(For ring final circuits see section 6.3.5.)

Every protective conductor, including the earthing conductor and main and supplementary bonding conductors, should be tested to verify that the conductors are electrically sound and correctly connected.

Test method 1 is suitable for testing the continuity of circuit protective conductors. Test method 2 is suitable for testing the continuity of earthing and bonding conductors and circuit protective conductors.

Test method 1 (for circuit protective conductors)

Test method 1, detailed below, in addition to providing for a check of the continuity of the protective conductor, also allows $(R_1 + R_2)$ to be measured, which, when added to the external impedance (Z_e), enables the earth fault loop impedance (Z_s) to be checked against the design value, see section 6.3.9.

NOTE: $(R_1 + R_2)$ is the sum of the resistances of the line conductor (R_1) and the circuit protective conductor (R_2) between the point of use and the origin of the installation.

Use an ohmmeter capable of measuring a low resistance for these tests.

Test method 1 can only be used to measure $(R_1 + R_2)$ for an 'all-insulated' installation. Installations incorporating steel conduit, steel trunking, micc (mineral insulated copper cables) and PVC/SWA cables will produce parallel paths to protective conductors. Such installations should be inspected for soundness of construction and test method 1 or 2 should be used to prove continuity.

▼ **Figure 6.3.4** Continuity of protective conductors using test method 1

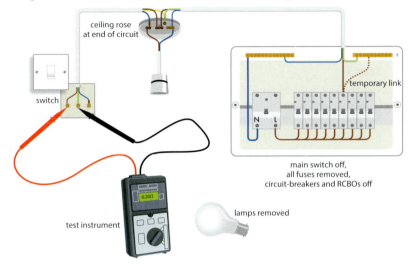

NOTE: The sheaths of cables entering the ceiling roses have been cut back for clarity in order to show the core colours. Sheaths must enclose the cable cores except within accessories.

Test method 1 procedure

If the instrument does not include an 'auto-null' facility, or it does but it is not used, the resistance of the test leads should be measured and deducted from the resistance readings obtained.

The line conductor to the circuit protective conductor should be bridged at the consumer unit or distribution board so as to include all the circuit. Thereafter, the line and earth terminals should be tested at each point in the circuit.

The measurement at the circuit's extremity should be recorded on the Schedule of Test Results Form 4 and is the value of $(R_1 + R_2)$ for the circuit being tested (see Figure 6.3.4).

Test method 2

Test method 2 may be used for checking the continuity of all protective conductors including earthing and bonding conductors. The method measures R_2 only.

Test method 2 procedure

Connect one terminal of the continuity test instrument to a long test lead and connect this to the consumer's main earthing terminal. Connect the other terminal of the instrument to another test lead and use this to make contact with the protective conductor at various points on the circuit, such as luminaires, switches, spur outlets, etc. The resistance of the protective conductor R_2 is recorded on the Schedule of Test Results, Form 4.

6.3.5 Continuity of ring final circuit conductors

A three-step test is required to verify the continuity of the line, neutral and protective conductors and correct wiring of a ring final circuit. The test results show if the ring has been correctly connected in an unbroken loop without interconnections.

Step 1

The line, neutral and protective conductors are identified at the distribution board or consumer unit and the end-to-end resistance of each is measured separately (see Figure 6.3.5a). These resistances are r_1, r_n and r_2 respectively. A finite reading confirms that there is no open circuit on the ring conductors under test. The resistance values obtained should be the same (within 0.05 ohm) if the conductors are all of the same size. If the protective conductor has a reduced csa the resistance r_2 of the protective conductor loop will be proportionally higher than that of the line and neutral loops, e.g. 1.67 times for 2.5/1.5 mm^2 cable. If these relationships are not achieved then either the conductors are incorrectly identified or there is something wrong at one or more of the accessories. The values of r_1, r_n and r_2 are recorded on the Schedule of Test Results.

Step 2

The line and neutral conductors are then connected together so that the outgoing line conductor is connected to the returning neutral conductor and vice versa (see Figure 6.3.5b). The resistance between line and neutral conductors is measured at each socket-outlet. The readings at each of the accessories wired into the ring will be substantially the same and the value will be approximately one-quarter of the resistance of the line plus the neutral loop resistances, i.e. $(r_1 + r_n)/4$. Any socket-outlets wired as spurs will have a higher resistance value due to the resistance of the spur conductors.

Step 3

The above step is then repeated, this time with the line and cpc cross-connected (see Figure 6.3.5c). The resistance between line and earth is measured at each socket-outlet. The readings obtained at each of the socket-outlets wired into the ring will be substantially the same and the value will be approximately one-quarter of the resistance of the line plus cpc loop resistances, i.e. $(r_1 + r_2)/4$. As before, a higher resistance value will be recorded at any socket-outlet wired as a spur. The highest value recorded represents the maximum $(R_1 + R_2)$ of the circuit and is recorded on the Schedule of Test Results. The value can be used to determine the earth loop impedance (Z_s) of the circuit to verify compliance with the loop impedance requirements of BS 7671 (see the Circuit Tables in chapter 4).

This sequence of tests also verifies the polarity at each socket-outlet and accessory unless the testing has been carried out at the terminals on the reverse of the socket-outlet. In such cases, a visual inspection of the connections is required to confirm correct polarity and dispense with the need for a separate polarity test.

▼ Figure 6.3.5a Initial check for continuity at ends of ring

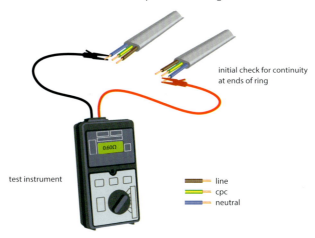

initial check for continuity
at ends of ring

test instrument

0.60Ω

▬▬ line
▬▬ cpc
▬▬ neutral

▼ Figure 6.3.5b

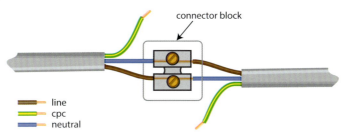

connector block

▬▬ line
▬▬ cpc
▬▬ neutral

▼ Figure 6.3.5c

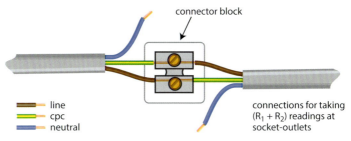

connector block

▬▬ line
▬▬ cpc
▬▬ neutral

connections for taking
$(R_1 + R_2)$ readings at
socket-outlets

NOTE: Where single-core cables are used, care should be taken to verify that the line and neutral conductors of opposite ends of the ring circuit are connected together. An error in this respect will be apparent from the readings taken at the socket-outlets. The readings will progressively increase in value towards the midpoint of the ring, then decrease again towards the other end of the ring.

6.3.6　Insulation resistance

Pre-test checks

(a) Check that the pilot or indicator lamps and capacitors are disconnected from circuits to prevent misleading test values from being obtained.

(b) If a circuit includes voltage-sensitive electronic devices such as dimmer switches, touch switches, delay timers, power controllers, electronic starters, controlgear for fluorescent lamps or RCDs with electronic amplifiers, etc., either:

- the devices must be temporarily disconnected; or
- a measurement should only be made between live conductors (line and neutral) connected together and the protective earth.

Tests

Tests should be carried out using the appropriate d.c. test voltage specified in Table 6.3.6.

The tests should be made at the consumer unit or distribution board with the main switch off, all fuses in place, switches and circuit-breakers closed, lamps removed and other current-using equipment disconnected. Earthing and equipotential bonding conductors are connected.

Where the removal of lamps and/or the disconnection of current-using equipment is impracticable, the local switches controlling such lamps and/or equipment should be open.

Where a circuit contains two-way switching the two-way switches must be operated one at a time and further insulation resistance tests carried out to ensure that all the circuit wiring is tested.

Although an insulation resistance value of not less than 1.0 megohm complies with BS 7671, where an insulation resistance of less than 2 megohms is recorded the possibility of a latent defect exists, this should be investigated.

Where electronic devices are disconnected for the purpose of the tests on the installation wiring (and the devices have exposed-conductive-parts that are required by BS 7671 to be connected to the protective conductors) the insulation resistance between the exposed-conductive-parts and all live parts of the device (line and neutral connected together) should be measured separately and should be not less than the values stated in Table 6.3.6.

▼ **Table 6.3.6** Minimum values of insulation resistance

Circuit nominal voltage	Test voltage (V d.c.)	Minimum insulation resistance (MΩ)
SELV and PELV	250	0.5
Up to and including 500 V with the exception of SELV and PELV	500	1.0

Insulation resistance between live conductors

Single-phase and three-phase

Test between all the live (line and neutral) conductors at the distribution board (see Figure 6.3.6a).

Resistance readings obtained should be not less than the minimum value stated in Table 6.3.6.

Insulation resistance to earth

Single-phase

Test between the live conductors (line and neutral) and the circuit protective conductors (connected to the earthing system) at the distribution board (see Figure 6.3.6b which illustrates neutral to earth only).

For a circuit containing two-way switching or two-way and intermediate switching, the switches must be operated one at a time and the circuit subjected to additional insulation resistance tests.

Three-phase

Test to earth from all live conductors (including the neutral) connected together. Where a low reading is obtained, it is necessary to test each conductor separately to earth, after disconnecting all equipment.

Resistance readings obtained should be not less than the minimum value stated to in Table 6.3.6.

NOTE: For further information on the measurement of earth fault loop impedance, refer to Guidance Note 3: *Inspection & Testing*.

SELV and PELV circuits

Test between SELV and PELV circuits and live parts of other circuits at 500 V d.c.

Test between SELV or PELV conductors at 250 V d.c. and between PELV conductors and protective conductors of the PELV circuit at 250 V d.c.

FELV circuits

FELV circuits are tested as low voltage circuits at 500 V d.c.

▼ **Figure 6.3.6a** Insulation resistance test between live conductors of a circuit

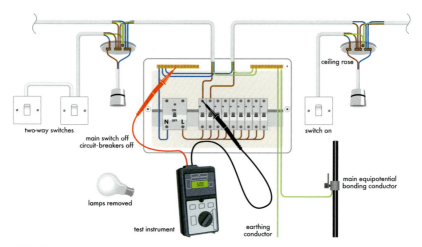

NOTES:

(a) The test should initially be carried out on the complete installation.

(b) The sheaths of cables entering the ceiling roses have been cut back for clarity in order to show the core colours. Sheaths must enclose the cable cores except within accessories.

▼ **Figure 6.3.6b** Insulation resistance test between neutral and earth

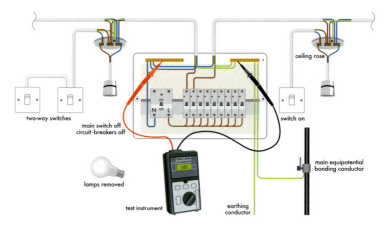

NOTES:

(a) The test should initially be carried out on the complete installation.

(b) Bonding and earthing connections are in place.

(c) The sheaths of cables entering the ceiling roses have been cut back for clarity in order to show the core colours. Sheaths must enclose the cable cores except within accessories.

6.3.7 Polarity

See Figure 6.3.7.

The method of test prior to connecting the supply is the same as test method 1 for checking the continuity of protective conductors, which should have already been carried out (see section 6.3.4). For ring final circuits, a visual check may be required (see 6.3.5 following step 3).

It is important to confirm that:

    (a) overcurrent devices and single-pole controls are connected in the line conductor;

    (b) except for E14 and E27 lampholders to BS EN 60238, centre contact screw lampholders have the outer threaded contact connected to the neutral; and

    (c) socket-outlet and similar accessory polarities are correct.

After connection of the supply, polarity must be checked using an approved voltage indicator or a test lamp (in either case with fused leads, see 6.3.1, GS 38).

▼ **Figure 6.3.7**　Polarity test on a lighting circuit

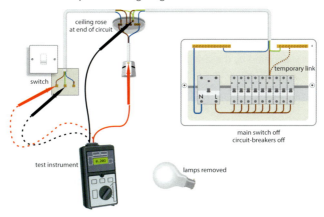

Note: the test may be carried out either at lighting points or switches

NOTES:

(a) The polarity of E14 and E27 Edison screw lampholders to BS EN 60238 does not have to be verified.

(b) The sheaths of cables entering the ceiling rose have been cut back for clarity in order to show the core colours. Sheaths must enclose the cable cores except within accessories.

(c) Polarity checks must be carried out on all circuits, i.e. power and lighting.

6.3.8 Earth electrode resistance

If the electrode under test is being used in conjunction with an RCD protecting an installation that forms part of a TT system, the following method of test may be applied.

With the main switch isolated and secured in the off position, a loop impedance tester is connected between the line conductor at the origin of the installation and the earth electrode with the test link open, and a test performed. This impedance reading is treated as the electrode resistance and is then added to the resistance of the protective conductor for the protected circuits. The test should be carried out before energizing the remainder of the installation.

The measured resistance should meet the following criteria and those of 6.3.9 but in any case should not exceed 200 ohms.

For TT systems, the value of the earth electrode resistance R_A in ohms multiplied by the operating current in amperes of the protective device $I_{\Delta n}$ should not exceed 50 V, e.g. if $R_A = 200 \ \Omega$, the maximum RCD operating current should not exceed 250 mA.

Remember to replace the test link.

6.3.9   Earth fault loop impedance

The earth fault loop impedance (Z_s) is required to be determined for the furthest point of each circuit. It may be determined by:

- direct measurement of Z_s; or
- direct measurement of Z_e at the origin and adding ($R_1 + R_2$) measured during the continuity tests (6.3.4 and 6.3.5) ($Z_s = Z_e + (R_1 + R_2)$); or
- adding ($R_1 + R_2$) measured during the continuity tests to the value of Z_e declared by the distributor (see section 3.2 or 4.1; e.g. 0.35 ohm for PME). The effectiveness of the distributor's earth must be confirmed by a test.

The external impedance (Z_e) may be measured using a line-earth loop impedance tester.

The main switch is opened and made secure to isolate the installation from the source of supply. The earthing conductor is disconnected from the Main Earthing Terminal and the measurement made between line and earth of the supply.

Remember to reconnect the earthing conductor to the earth terminal after the tests.

Direct measurement of Z_s can only be made on a live installation. Neither the connection with earth nor the bonding conductors are disconnected. The reading given by the loop impedance tester will usually be less than $Z_e + (R_1 + R_2)$ because of parallel earth return paths provided by any bonded extraneous-conductive-parts. This must be taken into account when comparing the results with design data.

Care should be taken to avoid any shock hazard to the testing personnel and to other persons on site during the tests.

The value of Z_s determined for each circuit should be less than the value given in the standard circuit schedules in Chapter 4 for the particular overcurrent device and cable.

For TN systems, when protection is afforded by an RCD, the rated residual operating current in amperes multiplied by the earth fault loop impedance in ohms should not exceed 50 V. This test should be carried out before energizing other parts of the system.

6.3.10 Measurement of prospective fault current

Installation designs should not be based on measured values of prospective fault current, as changes to the distribution network subsequent to the completion of the installation may increase fault levels.

Installation designs should be based on the maximum fault current provided by the distributor for a domestic premises, normally a conditional rating of 16 kA (see Chapter 3).

If it is desired to measure prospective fault levels, this should be done with all main bonding in place. Measurements are made at the distribution board between live conductors and between line conductors and earth.

For three-phase supplies, the maximum possible fault level will be approximately twice the single-phase to neutral value.

6.3.11 Functional testing

RCDs should be tested as described in section 6.4. All assemblies including switchgear, controls and interlocks should be functionally tested; that is, operated to check that they work and are properly fixed etc.

6.4 Operation of residual current devices (RCDs) and residual current breakers with overcurrent protection (RCBOs)

6.4.1 Test procedure

Where a residual current device (RCD) provides fault protection or additional protection, its effectiveness must be verified by a test simulating an appropriate fault condition and independent of any test facility incorporated in the device. While the fault may be simulated by a simple test, reliable operation of the RCD is best ensured by using an RCD tester and applying the range of tests described below. Test results are recorded on the test results schedule.

The tests are made on the load side of the RCD, as near as practicable to its point of installation, and between the line conductor of the protected circuit and the associated circuit protective conductor. The load supplied should be disconnected during the tests.

The tests can produce potentially dangerous voltages on exposed-conductive-parts and extraneous-conductive-parts when the loop impedance of the circuit approaches the maximum limit, so precautions must therefore be taken to prevent persons or livestock touching such parts.

6.4.2 General-purpose RCDs to BS 4293

(a) With a leakage current flowing equivalent to 50 per cent of the rated tripping current $I_{\Delta n}$ of the RCD, the device should not open.

(b) With a leakage current flowing equivalent to 100 per cent of the rated tripping current $I_{\Delta n}$ of the RCD, the device should open in less than 200 ms. Where the RCD incorporates an intentional time delay, it should trip within a time range from '50 per cent of the rated time delay plus 200 ms' to '100 per cent of the rated time delay plus 200 ms'.

6.4.3 General-purpose RCCBs to BS EN 61008 or RCBOs to BS EN 61009

(a) With a leakage current flowing equivalent to 50 per cent of the rated tripping current $I_{\Delta n}$ of the RCD, the device should not open.

(b) With a leakage current flowing equivalent to 100 per cent of the rated tripping current $I_{\Delta n}$ of the RCD, the device should open in less than 300 ms unless it is of 'Type S' (or selective), which incorporates an intentional time delay. In this case, it should trip within a time range from 130 ms to 500 ms.

6.4.4 RCD protected socket-outlets to BS 7288

(a) With a leakage current flowing equivalent to 50 per cent of the rated tripping current $I_{\Delta n}$ of the RCD, the device should not open.

(b) With a leakage current flowing equivalent to 100 per cent of the rated tripping current $I_{\Delta n}$ of the RCD, the device should open in less than 200 ms.

(c) With a leakage current flowing equivalent to 5 times the rated tripping current $I_{\Delta n}$ of the RCD, the device should open in less than 40 ms

6.4.5 Additional protection

Where an RCD or RCBO with a rated residual operating current $I_{\Delta n}$ not exceeding 30 mA is used to provide additional protection (against direct contact with energized parts), with a test current of $5I_{\Delta n}$, the device should open in less than 40 ms. The maximum test time must not be longer than 40 ms, unless the protective conductor potential rises by less than 50 V. (The instrument supplier will advise on compliance.)

6.4.6 Integral test device

An integral test device is incorporated in each RCD. This device enables the electrical and mechanical parts of the RCD to be verified by pressing the button marked 'T' or 'Test'. Operation of the integral test device does not provide a means of checking:

(a) the continuity of the earthing conductor or the associated circuit protective conductors; or

(b) any earth electrode or other means of earthing; or

(c) any other part of the associated installation earthing.

The test button will only operate the RCD if the device is energized. Confirm that the notice to test RCDs quarterly (by pressing the test button) is fixed in a prominent position (see section 3.4.8).

6.4.7 Test failure

If any test shows a failure to comply, the installation fault must be corrected. The test must then be repeated, as must any earlier test that could have been influenced by the failure.

Safe working 7

7.1 Safety policy and risk assessment

The Electrotechnical Assessment Scheme for the technical competence of enterprises that undertake electrical installation work in dwellings requires that:

'The enterprise shall have a written health and safety policy statement and will carry out risk assessments as appropriate.'

The Management of Health and Safety at Work Regulations 1999 require every employer to make a suitable and sufficient assessment of the risks to the health and safety of their employees to which they are exposed while at work; and the risks to health and safety if persons not in their direct employment arising out of, or in connection with, the work being undertaken. The Health and Safety Executive has issued an Approved Code of Practice – L21, to provide guidance on these risks. It is likely that all but the smallest of enterprises will need a copy of the approved code of practice and will need to prepare a written health and safety policy. Not only are industrial and commercial premises places of work, but dwellings are also places of work during construction, maintenance, repair and addition and alteration. The management of the electrical installation enterprise have responsibilities for their own staff and for others working on the site. It is important that all employees understand the commitment to the safety of the enterprise and know the procedures that they are expected to follow.

The following procedures for assessing the condition of the installation to be worked upon and the procedures to be followed for working are intended to help the managers of an enterprise in advising employees of the safety procedures to be followed. Typical safety procedures, including both non-electrical as well as electrical safety instructions and permit-to-work systems, can be found in Appendix D.

7.2 Pre-work survey

Additions and alterations to existing electrical installations

A number of checks must be carried out on an existing installation before any changes are made, as described in paragraph 1.7.

Approved Document P in Section 1 *Design and installation* states:

New dwellings formed by a change of use

1.5 Where a material change of use creates a new dwelling, or changes the number of dwellings in a building, regulation 6 requires that any necessary work is carried out to ensure that the building complies with requirement P1. This means that in some cases the existing electrical installation will need to be upgraded to meet current standards.

NOTE: If existing cables are adequate, it is not necessary to replace them, even if they use old colour codes.

Additions and alterations to existing electrical installations

1.6 Regulation 4(3) states that when building work is complete, the building should be no more unsatisfactory in terms of complying with the applicable parts of Schedule 1 to the Building Regulations than before the building work was started. Therefore, when extending or altering an electrical installation, only the new work must meet current standards. There is no obligation to upgrade the existing installation unless either of the following applies:

(a) The new work adversely affects the safety of the existing installation.

(b) The state of the existing installation is such that the new work cannot be operated safely.

1.7 Any new work should be carried out in accordance with BS 7671. The existing electrical installation should be checked to ensure that the following conditions are all satisfied.

(a) The rating and condition of existing equipment belonging to both the consumer and to the electricity distributor are suitable for the equipment to carry the additional loads arising from the new work.

(b) Adequate protective measures are used.

(c) The earthing and equipotential bonding arrangements are satisfactory.

7.3   Pre-work tests

Electricians should carry out the following tests before any installation work is started, including additions and alterations to existing installations, rewires and new installations (where applicable):

▶ polarity;
▶ effectiveness of earthing; and
▶ operation of residual current devices (RCDs).

Any deficiencies in the polarity or earthing arrangements must be rectified before work on the installation is commenced.

The procedures require all work other than certain specified tests to be carried out on dead electrical equipment. The Electricity at Work Regulations 1989 prohibit employees working on or near live equipment unless in all the circumstances it is unreasonable for it to be switched off. In domestic premises it is unlikely that there will be circumstances where it is unreasonable to make the equipment dead before working on it.

Before commencing work:

(a) Check that it is safe and acceptable to the persons in the premises to switch off the supply.
(b) Identify the electrical equipment to be worked on and the means of disconnection from all points of supply, e.g. by the opening of circuit-breakers, isolating switches, removal of fuses, links, or other suitable means. For a domestic installation, unless unreasonable in the circumstances, the main switch is opened and secured if necessary, making the complete electrical installation dead.
(c) Isolate and prove. An approved test instrument, e.g. test lamp or voltage indicator, is used to verify that the installation or the part of the installation to be worked on is dead. The Electricity at Work Regulations do not allow live working (unless it is reasonable in the circumstances to work with live conditions). The procedure for isolating and proving is:
  (i) prove the operation of the test instrument by receiving positive indication that the instrument works, e.g. lamp glows with proving unit. The preferred proof of operation of the device is a live test of the circuit to be isolated, if this is practical and safe.
  (ii) isolate the installation and secure the means of isolation.
  (iii) test all the conductors of the isolated circuits for voltage to earth with the test instrument. It should be confirmed that protective conductors are not live because of, for example, a wiring fault.
  (iv) prove the operation of the test instrument again by receiving positive indication that the tester works, e.g. lamp glows with proving unit.
  (v) if necessary for reasons of safety, earth the line conductor with care, treating it as live.
(d) Warning notices may need to be placed against interference.
(e) If there are persons in the property, they must be advised that they must **under no circumstances attempt to switch on the isolating device.**

7.4   Isolation

Regulation 12 (1) (b) of the Electricity at Work Regulations 1989 requires that there will be a suitable means of ensuring that a supply that has been switched off will remain switched off and inadvertent reconnection prevented. Consequently, precautions need to be taken to prevent electrical equipment being made live inadvertently by others in the premises, particularly other electricians. This may be achieved by:

(a) an approved lock;

(b) removal of fuses or links; or

(c) in the case of equipment supplied via a plug and socket, removal of the plug from the socket-outlet. If the socket-outlet is not in view of the person carrying out the work, a suitable label warning against interference should be fixed or a lockable box employed.

As indicated above, the electrical equipment should be proved dead by the proper use of an approved voltage indicating device. Alternatively, clear evidence of isolation is needed, such as the physical tracing of the circuit and physical identification of the means of isolation, e.g. removal of the plug. On no account should reliance be placed on time switches, limit switches, lockout push buttons, etc.

NOTE: Electricity distribution company procedures for removal and replacement of electricity company cut-out fuses must be followed. Permission from the company is required.

A typical isolation procedure is shown in the diagram opposite.

7.5   Defects in electrical installations

Defects in an electrical installation that are identified before or during the work that affect the addition or alterations being carried out must be corrected before the work is completed, made live and certificates issued. Such work would include provision of adequate earthing arrangements and the installation of suitable residual current devices.

Defects identified in an installation that do not affect the addition or alteration being carried out are required to be notified in writing to the householder. The notification should be confirmed on the completion certificate.

Twelve easy steps to safe isolation

17th EDITION

NAPIT

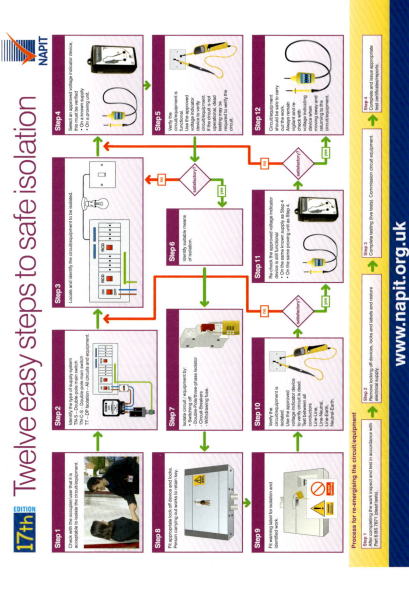

Step 1
Check with the occupier/user that it is acceptable to isolate the circuit/equipment.

Step 2
Identify the type of supply system
TN-S – Double-pole main switch
TN-C-S – Double-pole main switch
TT - DP Isolation - All circuits and equipment.

Step 3
Locate and identify the circuit/equipment to be isolated.

Step 4
Select an approved voltage indicator device, this must be verified
• On a known supply
• On a proving unit.

Step 5
Verify the circuit/equipment is functional. Use the approved voltage indicator device to verify circuit/equipment. If the circuit is not operational, dead testing may be required to verify the circuit.

satisfactory? no / yes

Step 6
Identify suitable means of isolation.

Step 7
Isolate circuit / equipment by;
• Switching off
- Double-Pole/three-phase Isolator
- Circuit-Breakers
- Withdrawing fuse.
Test between all conductors
Line-Line,
Line-Neutral,
Line-Earth,
Neutral-Earth.

Step 8
Fit appropriate lock off device and locks. Person carrying out works to retain key.

Step 9
Fit warning label for isolation and identified work.

Step 10
Verify the circuit/equipment is isolated. Use the approved voltage indicator device to verify circuit is dead. Test between all conductors Line-Line, Line-Neutral, Line-Earth, Neutral-Earth.

satisfactory? no / yes

Step 11
Re-check the approved voltage indicator device is still functional
• On the same known supply as Step 4
• On the same proving unit as Step 4.

satisfactory? no / yes

Step 12
Circuit/equipment should be safe to carry out the work. Always remain vigilant and re-check with voltage indicating device when moving away and returning to the circuit/equipment.

Process for re-energising the circuit/equipment

Step 1
After completing the work inspect and test in accordance with Part 6 BS 7671 (dead tests).

Step 2
Remove locking off devices, locks and labels and restore electrical supply.

Step 3
Complete testing (live tests). Commission circuit equipment.

Step 4
Complete and issue appropriate test certificates/reports.

www.napit.org.uk

Maintenance 8

8.1 The need for maintenance

Regulation 4(2) of the Electricity at Work Regulations 1989 requires that:

As may be necessary to prevent danger, all systems shall be maintained so as to prevent, so far as is reasonably practicable, such danger.

In the Electricity at Work Regulations systems are defined as follows:

> *'system' means an electrical system in which all the electrical equipment is, or may be, electrically connected to a common source of electrical energy, and includes such source and such equipment.* (Regulation 2(1).)

where, also defined:

> *'electrical equipment' includes anything used, intended to be used or installed for use, to generate, provide, transmit, transform, rectify, convert, conduct, distribute, control, store, measure or use electrical energy.* (Regulation 2(1).)

It is clear that, as a consequence of these definitions, 'system' includes all electrical equipment including the generating equipment, the fixed wiring of a building, and all the equipment in the building, including fixed, mobile and hand-held appliances. Electrical equipment includes anything powered by any source of electrical energy, including battery-powered.

8.2 Domestic installations

Domestic installations are within the compass of the Electricity at Work Regulations when persons are at work, i.e. employed to work in the premises. However, the same basic principles apply as are appropriate to places of work, in that maintenance of domestic installations would comprise:

(a) routine checks; and
(b) periodic inspection and, as necessary, testing.

In a user manual, a householder must be advised that, as well as a professional periodic inspection and test at least every ten years, all defects must be repaired as they arise. If there are signs of wear, overheating, looseness, or difficulty in operating equipment then an electrician should be instructed to inspect the installation and maintain as necessary.

8.3 Purpose of periodic inspection (condition reporting)

The purpose of periodic inspection and testing of an electrical installation is to determine, so far as is reasonably practicable, whether the installation is in a satisfactory condition for continued use. This does not necessarily mean compliance with all the latest requirements of BS 7671.

BS 7671 requires an inspection comprising a detailed examination of the installation to be carried out without dismantling or with partial dismantling as required, together with appropriate tests. The purpose of the inspection and testing as stated in BS 7671 is to provide so far as reasonably practicable for:

(a) the safety of persons and livestock against the effects of electric shock and burns;

(b) protection against damage to property by fire and heat arising from installation defects;

(c) confirmation that the installation is not damaged or deteriorated so as to impair safety; and

(d) the identification of installation defects and non-compliance with the BS 7671 that may give rise to danger.

It is to be noted that the essence of periodic inspection is to identify those defects that may give rise to danger.

8.4 Electrical Installation Condition Report

Appendix 6 of BS 7671 includes an Electrical Installation Condition Report complete with notes for the inspector and for the recipient of the report. The following forms are as in BS 7671 except they have been completed for a 'typical' periodic inspection and test.

Electrical Installation Condition Report forms complete with inspection and test schedules may be downloaded from the Institution's website: www.theiet.org

8.4.1 Typical Electrical Installation Condition Report

▼ **Form 6** Electrical installation condition report (F6) (note 1)

ELECTRICAL INSTALLATION CONDITION REPORT Form 6 No. *126-2* /6

SECTION A. DETAILS OF THE CLIENT / PERSON ORDERING THE REPORT
Name *Mr A Brown*
Address *111 Any Street*
TOWN Postcode: *NT7 8SR*

SECTION B. REASON FOR PRODUCING THIS REPORT *Mortgage*

Date(s) on which inspection and testing was carried out

SECTION C. DETAILS OF THE INSTALLATION WHICH IS THE SUBJECT OF THIS REPORT
Occupier *As above*
Address *As Above*
Postcode:

Description of premises
Domestic ☑ Commercial ☐ Industrial ☐ Other (include brief description) ☐
Estimated age of wiring system *15* years
Evidence of additions/alterations Yes ☑ No ☐ Not apparent ☑ If yes, estimate age *5* years
Installation records available? (Regulation 621.1) Yes ☐ No ☑ Date of last inspection (date)

SECTION D. EXTENT AND LIMITATIONS OF INSPECTION AND TESTING
Extent of the electrical installation covered by this report
Installation to house and garage

Agreed limitations including the reasons (see Regulation 634.2) *No dismantling, lifting of floor coverings*

Agreed with: *Mr Brown*
Operational limitations including the reasons (see page no............) *None*

The inspection and testing detailed in this report and accompanying schedules have been carried out in accordance with BS 7671: 2008 (IET Wiring Regulations) as amended to *2015*.
It should be noted that cables concealed within trunking and conduits, under floors, in roof spaces, and generally within the fabric of the building or underground, have **not** been inspected unless specifically agreed between the client and inspector prior to the inspection. An inspection should be made within an accessible roof space housing other electrical equipment.

SECTION E. SUMMARY OF THE CONDITION OF THE INSTALLATION
General condition of the installation (in terms of electrical safety) *Fixed wiring in good condition, lighting pendants require replacing and RCDs required.*

Overall assessment of the installation in terms of its suitability for continued use
~~SATISFACTORY~~ / UNSATISFACTORY* (Delete as appropriate)
*An unsatisfactory assessment indicates that dangerous (code C1) and/or potentially dangerous (code C2) conditions have been identified.

SECTION F. RECOMMENDATIONS
Where the overall assessment of the suitability of the installation for continued use above is stated as UNSATISFACTORY, I / we recommend that any observations classified as 'Danger present' (code C1) or 'Potentially dangerous' (code C2) are acted upon as a matter of urgency.
Investigation without delay is recommended for observations identified as 'Further investigation required' (code FI).
Observations classified as 'Improvement recommended' (code C3) should be given due consideration.

Subject to the necessary remedial action being taken, I/we recommend that the installation is further inspected and tested by *July 2018* (date)

SECTION G. DECLARATION
I/We, being the person(s) responsible for the inspection and testing of the electrical installation (as indicated by my/our signatures below), particulars of which are described above, having exercised reasonable skill and care when carrying out the inspection and testing, hereby declare that the information in this report, including the observations and the attached schedules, provides an accurate assessment of the condition of the electrical installation taking into account the stated extent and limitations in section D of this report.

Inspected and tested by:	Report authorised for issue by:
Name (Capitals) *W WHITE*	Name (Capitals) *A BOSS*
Signature *W White*	Signature *A Boss*
For/on behalf of *County Electrics*	For/on behalf of *County Electrics*
Position *Electrician*	Position *Qualified Supervisor*
Address *187 Industrial Lane, OLDTOWN*	Address *187 Industrial Lane, OLDTOWN*
Date *18-July-2015* Postcode: *NT7 1ST*	Date *21-July-2015* Postcode: *NT7 1ST*

SECTION H. SCHEDULE(S)
1 schedule(s) of inspection and *1* schedule(s) of test results are attached.
The attached schedule(s) are part of this document and this report is valid only when they are attached to it.

Page 1 of *5*

▼ **Form 6** *continued*

SECTION I. SUPPLY CHARACTERISTICS AND EARTHING ARRANGEMENTS

Earthing arrangements	Number and Type of Live Conductors		Nature of Supply Parameters	Supply Protective Device
TN-C ☐	a.c. ☑	d.c. ☐	Nominal voltage, U / U₀$^{(1)}$ .....230... V	BS (EN) ....1361....
TN-S ☐	1-phase, 2-wire ☑	2-wire ☐	Nominal frequency, f$^{(1)}$ .....50... Hz	Type ......type II....
TN-C-S ☑	2 phase, 3-wire ☐	3-wire ☐	Prospective fault current, I$_{pf}^{(2)}$ .....1.0.. kA	Rated current .....100..A
TT ☐	3 phase, 3-wire ☐	Other ☐	External loop impedance, Ze$^{(2)}$ ...0.24.. Ω	
IT ☐	3 phase, 4-wire ☐		(Note: (1) by enquiry (2) by enquiry or by measurement)	
	Confirmation of supply polarity ☑			

Other sources of supply (as detailed on attached schedule) ☐

SECTION J. PARTICULARS OF INSTALLATION REFERRED TO IN THE REPORT

Means of Earthing | **Details of Installation Earth Electrode** *(where applicable)*

Distributor's facility ☑   Type .....*Not applicable*.....

Installation earth electrode ☐   Location ....

Resistance to Earth ....Ω

Main Protective Conductors

Earthing conductor	Material ..*Copper*.. csa ......10....mm²	Connection / continuity verified ☑
Main protective bonding conductors (to extraneous-conductive-parts)	Material ..*Copper*.. csa ......6....mm²	Connection / continuity verified ☑

To water installation pipes ☑   To gas installation pipes ☑   To oil installation pipes ☐   To structural steel ☐

To lightning protection ☐   To other ☐   Specify ....

Main Switch / Switch-Fuse / Circuit-Breaker / RCD

Location ..*Under stairs*..	Current rating .....80...A	**If RCD main switch**
....	Fuse / device rating or setting *N/A* A	Rated residual operating current (I$_{\Delta n}$) ..*N/A*...mA
BS(EN) ....5486/2..	Voltage rating .....240..V	Rated time delay ....ms
No of poles ....2..		Measured operating time(at I$_{\Delta n}$) ....ms

SECTION K. OBSERVATIONS

Referring to the attached schedules of inspection and test results, and subject to the limitations specified at the *Extent and limitations of inspection and testing* section

No remedial action is required ☐   The following observations are made ☑ (see below):

OBSERVATION(S) Include schedule reference, as appropriate	CLASSIFICATION CODE
1. Broken socket with exposed live parts in kitchen	C1
2. Lighting pendants signs of overheating	C2
3. No RCD to sockets	C3
4. No RCD to bathroom sockets	C3
5. Main bonding conductors 6 mm², should be 10 mm²	C3
6. Consumer unit not labelled	C3
NOTE: Consumer unit plastic, not metal (whilst not a requirement at the moment, after Jan 2016 it will become a requirement)	

One of the following codes, as appropriate, has been allocated to each of the observations made above to indicate to the person(s) responsible for the installation the degree of urgency for remedial action.

C1 – Danger present. Risk of injury. Immediate remedial action required

C2 – Potentially dangerous - urgent remedial action required

C3 – Improvement recommended

FI – Further investigation required without delay

CONDITION REPORT

Notes for the person producing the Report:

1 This Report should only be used for reporting on the condition of an existing electrical installation. An installation which was designed to an earlier edition of the Regulations and which does not fully comply with the current edition is not necessarily unsafe for continued use, or requires upgrading. Only damage, deterioration, defects, dangerous conditions and non-compliance with the requirements of the Regulations, which may give rise to danger, should be recorded.

2 The Report, normally comprising at least five pages, should include schedules of both the inspection and the test results. Additional pages may be necessary for other than a simple installation and for the 'Guidance for recipients'. The number of each page should be indicated, together with the total number of pages involved.

3 The reason for producing this Report, such as change of occupancy or landlord's periodic maintenance, should be identified in Section B.

4 Those elements of the installation that are covered by the Report and those that are not should be identified in Section D (Extent and limitations). These aspects should have been agreed with the person ordering the Report and other interested parties before the inspection and testing commenced. Any operational limitations, such as inability to gain access to parts of the installation or an item of equipment, should also be recorded in Section D.

5 The maximum prospective value of fault current (I_{pf}) recorded should be the greater of either the prospective value of short-circuit current or the prospective value of earth fault current.

6 Where an installation has an alternative source of supply a further schedule of supply characteristics and earthing arrangements based upon Section I of this Report should be provided.

7 A summary of the condition of the installation in terms of safety should be clearly stated in Section E. Observations, if any, should be categorised in Section K using the coding C1 to C3 as appropriate. Any observation given a code C1 or C2 classification should result in the overall condition of the installation being reported as unsatisfactory.

8 Wherever practicable, items classified as 'Danger present' (C1) should be made safe on discovery. Where this is not possible the owner or user should be given written notification as a matter of urgency.

9 Where an observation requires further investigation (FI) because the inspection has revealed an apparent deficiency which could not, owing to the extent or limitations of the inspection, be fully identified, and further investigation may reveal a code C1 or C2 item, this should be recorded within Section K.

10 If the space available for observations in Section K is insufficient, additional pages should be provided as necessary.

11 The date by which the next Electrical Installation Condition Report is recommended should be given in Section F. The interval between inspections should take into account the type and usage of the installation and its overall condition.

CONDITION REPORT

GUIDANCE FOR RECIPIENTS (to be appended to the Report)

This Report is an important and valuable document, which should be retained for future reference.

1 The purpose of this Condition Report is to confirm, so far as reasonably practicable, whether or not the electrical installation is in a satisfactory condition for continued service (see Section E). The Report should identify any damage, deterioration, defects and/or conditions, which may give rise to danger (see Section K).

2 The person ordering the Report should have received the "original" Report and the inspector should have retained a duplicate.

3 The "original" Report should be retained in a safe place and be made available to any person inspecting or undertaking work on the electrical installation in the future. If the property is vacated, this Report will provide the new owner /occupier with details of the condition of the electrical installation at the time the Report was issued.

4 Where the installation incorporates a residual current device (RCD) there should be a notice at or near the device stating that it should be tested quarterly. For safety reasons it is important that this instruction is followed.

5 Section D (Extent and Limitations) should identify fully the extent of the installation covered by this Report and any limitations on the inspection and testing. The inspector should have agreed these aspects with the person ordering the Report and with other interested parties (licensing authority, insurance company, mortgage provider and the like) before the inspection was carried out.

6 Some operational limitations such as inability to gain access to parts of the installation or an item of equipment may have been encountered during the inspection. The inspector should have noted these in Section D.

7 For items classified in Section K as C1 ("Danger present"), the safety of those using the installation is at risk, and it is recommended that a electrically skilled person undertakes the necessary remedial work immediately.

8 For items classified in Section K as C2 ("Potentially dangerous"), the safety of those using the installation may be at risk and it is recommended that a electrically skilled person undertakes the necessary remedial work as a matter of urgency.

9 Where it has been stated in Section K that an observation requires further investigation (code FI) the inspection has revealed an apparent deficiency which may result in a code C1 or C2, and could not, due to the extent or limitations of the inspection, be fully identified. Such observations should be investigated as soon as possible. A further examination of the installation will be necessary, to determine the nature and extent of the apparent deficiency (see Section F).

10 For safety reasons, the electrical installation should be re-inspected at appropriate intervals by a electrically skilled person. The recommended date by which the next inspection is due is stated in Section F of the Report

under 'Recommendations' and on a label at or near to the consumer unit / distribution board.

CONDITION REPORT INSPECTION SCHEDULE

GUIDANCE FOR THE INSPECTOR

1 Section 1.0. Where inadequacies in the distributor's equipment are encountered the inspector should advise the person ordering the work to inform the appropriate authority.
2 Older installations designed prior to BS 7671:2008 may not have been provided with RCDs for additional protection. The absence of such protection should as a minimum be given a code C3 classification (item 5.12).
3 The schedule is not exhaustive.
4 Numbers in brackets are Regulation references to specified requirements.

▼ Form 7   Condition Report Inspection Schedule

**CONDITION REPORT INSPECTION SCHEDULE FOR
DOMESTIC AND SIMILAR PREMISES WITH UP TO 100 A SUPPLY**

Form 7 No. ...... **126-2** /7

Note: This form is suitable for many types of smaller installation, not exclusively domestic.

OUTCOMES	Acceptable condition	✓	Unacceptable condition	State C1 or C2	Improvement recommended	State C3	Further investigation	FI	Not verified	N/V	Limitation	LIM	Not applicable	N/A

ITEM NO	DESCRIPTION	OUTCOME (Use codes above. Provide additional comment where appropriate. C1, C2, C3 and FI coded items to be recorded in Section K of the Condition Report)
1.0	**DISTRIBUTOR'S / SUPPLY INTAKE EQUIPMENT**	
1.1	Condition of service cable	✓
1.2	Condition of service head	✓
1.3	Condition of distributor's earthing arrangement	✓
1.4	Condition of meter tails – Distributor/Consumer	✓
1.5	Condition of metering equipment	✓
1.6	Condition of isolator (where present)	✓
2.0	**PRESENCE OF ADEQUATE ARRANGEMENTS FOR OTHER SOURCES SUCH AS MICROGENERATORS (551.6; 551.7)**	N/A
3.0	**EARTHING / BONDING ARRANGEMENTS (411.3; Chap 54)**	
3.1	Presence and condition of distributor's earthing arrangement (542.1.2.1; 542.1.2.2)	✓
3.2	Presence and condition of earth electrode connection where applicable (542.1.2.3)	N/A
3.3	Provision of earthing/bonding labels at all appropriate locations (514.13.1)	✓
3.4	Confirmation of earthing conductor size (542.3; 543.1.1)	✓
3.5	Accessibility and condition of earthing conductor at MET (543.3.2)	✓
3.6	Confirmation of main protective bonding conductor sizes (544.1)	✓
3.7	Condition and accessibility of main protective bonding conductor connections (543.3.2; 544.1.2)	✓
3.8	Accessibility and condition of other protective bonding connections (543.3.2)	✓
4.0	**CONSUMER UNIT(S) / DISTRIBUTION BOARD(S)**	
4.1	Adequacy of working space/accessibility to consumer unit/distribution board (132.12; 513.1)	✓
4.2	Security of fixing (134.1.1)	✓
4.3	Condition of enclosure(s) in terms of IP rating etc (416.2)	✓
4.4	Condition of enclosure(s) in terms of fire rating etc (421.1.201, 526.5)	✓
4.5	Enclosure not damaged/deteriorated so as to impair safety (621.2(iii))	✓
4.6	Presence of main linked switch (as required by 537.1.4)	✓
4.7	Operation of main switch (functional check) (612.13.2)	✓
4.8	Manual operation of circuit-breakers and RCDs to prove disconnection (612.13.2)	✓
4.9	Correct identification of circuit details and protective devices (514.8.1; 514.9.1)	✓
4.10	Presence of RCD quarterly test notice at or near consumer unit/distribution board (514.12.2)	C3
4.11	Presence of non-standard (mixed) cable colour warning notice at or near consumer unit/distribution board (514.14)	✓
4.12	Presence of alternative supply warning notice at or near consumer unit/distribution board (514.15)	N/A
4.13	Presence of other required labelling (please specify) (Section 514)	N/A
4.14	Examination of protective device(s) and base(s); correct type and rating (no signs of unacceptable thermal damage, arcing or overheating) (421.1.3)	✓
4.15	Single-pole switching or protective devices in line conductors only (132.14.1; 530.3.2)	✓
4.16	Protection against mechanical damage where cables enter consumer unit/distribution board (522.8.1; 522.8.11)	✓
4.17	Protection against electromagnetic effects where cables enter consumer unit/distribution board/enclosures (521.5.1)	✓
4.18	RCD(s) provided for fault protection – includes RCBOs (411.4.9; 411.5.2; 531.2)	C3
4.19	RCD(s) provided for additional protection - includes RCBOs (411.3.3; 415.1)	C3
4.20	Confirmation of indication that SPD is functional (534.2.8)	N/A
4.21	Confirmation that ALL conductor connections, including connections to busbars, are correctly located in terminals and are tight and secure (526.1)	✓
4.22	Adequate arrangements where a generating set operates as a switched alternative to the public supply (551.6)	N/A
4.23	Adequate arrangements where a generating set operates in parallel with the public supply (551.7)	N/A

Page 3.. of .5.

▼ **Form 7** Condition Report Inspection Schedule *continued*

Form 7 No. ......**126-2** /7

OUTCOMES	Acceptable condition	✓	Unacceptable condition	State C1 or C2	Improvement recommended	State C3	Further investigation	FI	Not verified	N/V	Limitation	LIM	Not applicable	N/A
ITEM NO			**DESCRIPTION**								**OUTCOME** *(Use codes above. Provide additional comment where appropriate. C1, C2, C3 and FI coded items to be recorded in Section K of the Condition Report)*			

ITEM NO	DESCRIPTION	OUTCOME
5.0	**FINAL CIRCUITS**	
5.1	Identification of conductors (514.3.1)	✓
5.2	Cables correctly supported throughout their run (522.8.5)	✓
5.3	Condition of insulation of live parts (416.1)	✓
5.4	Non-sheathed cables protected by enclosure in conduit, ducting or trunking (521.10.1)	✓
	▪ To include the integrity of conduit and trunking systems (metallic and plastic)	✓
5.5	Adequacy of cables for current-carrying capacity with regard for the type and nature of installation (Section 523)	✓
5.6	Coordination between conductors and overload protective devices (433.1; 533.2.1)	✓
5.7	Adequacy of protective devices: type and rated current for fault protection (411.3)	✓
5.8	Presence and adequacy of circuit protective conductors (411.3.1.1; 543.1)	✓
5.9	Wiring system(s) appropriate for the type and nature of the installation and external influences (Section 522)	✓
5.10	Concealed cables installed in prescribed zones (see Section D. *Extent and limitations*) (522.6.202)	N/A
5.11	Cables concealed under floors, above ceilings or in walls/partitions, adequately protected against damage (see Section D. *Extent and limitations*) (522.6.204)	N/A
5.12	Provision of additional protection by RCD not exceeding 30 mA:	
	▪ for all socket-outlets of rating 20 A or less, unless an exception is permitted (411.3.3)	C2
	▪ for supply to mobile equipment not exceeding 32 A rating for use outdoors (411.3.3)	C3
	▪ for cables concealed in walls at a depth of less than 50 mm (522.6.202, .203)	C3
	▪ for cables concealed in walls/partitions containing metal parts regardless of depth (522.6.203)	N/A, C1
5.13	Provision of fire barriers, sealing arrangements and protection against thermal effects (Section 527)	✓
5.14	Band II cables segregated/separated from Band I cables (528.1)	N/A
5.15	Cables segregated/separated from communications cabling (528.2)	N/A
5.16	Cables segregated/separated from non-electrical services (528.3)	✓
5.17	Termination of cables at enclosures – indicate extent of sampling in Section D of the report (Section 526)	
	▪ Connections soundly made and under no undue strain (526.6)	✓
	▪ No basic insulation of a conductor visible outside enclosure (526.8)	✓
	▪ Connections of live conductors adequately enclosed (526.5)	✓
	▪ Adequately connected at point of entry to enclosure (glands, bushes etc.) (522.8.5)	✓
5.18	Condition of accessories including socket-outlets, switches and joint boxes (621.2(iii))	C1
5.19	Suitability of accessories for external influences (512.2)	✓
5.20	Adequacy of working space/accessibility to equipment (132.12; 513.1)	✓
5.21	Single-pole switching or protective devices in line conductors only (132.14.1, 530.3.2)	✓
6.0	**LOCATION(S) CONTAINING A BATH OR SHOWER**	
6.1	Additional protection for all low voltage (LV) circuits by RCD not exceeding 30 mA (701.411.3.3)	C3
6.2	Where used as a protective measure, requirements for SELV or PELV met (701.414.4.5)	N/A
6.3	Shaver sockets comply with BS EN 61558-2-5 formerly BS 3535 (701.512.3)	N/A
6.4	Presence of supplementary bonding conductors, unless not required by BS 7671:2008 (701.415.2)	✓
6.5	Low voltage (e.g. 230 volt) socket-outlets sited at least 3 m from zone 1 (701.512.3)	✓
6.6	Suitability of equipment for external influences for installed location in terms of IP rating (701.512.2)	✓
6.7	Suitability of accessories and controlgear etc. for a particular zone (701.512.3)	✓
6.8	Suitability of current-using equipment for particular position within the location (701.55)	✓
7.0	**OTHER PART 7 SPECIAL INSTALLATIONS OR LOCATIONS**	
7.1	List all other special installations or locations present, if any. (Record separately the results of particular inspections applied.)	N/A

Inspected by:
Name (Capitals) ...W WHITE......   Signature ......*W White*......   Date ..18/7/2015..

Page ..4.. of ..5..

▶ Form 4　Schedule of Test Results

GENERIC SCHEDULE OF TEST RESULTS

Form 4 No. ...126-2...../4

DB reference no	Consumer unit
Location	Under stairs
Zs at DB (Ω)	0.24
Ipf at DB (kA)	1.0
Correct supply polarity confirmed	☑
Phase sequence confirmed (where appropriate)	☐

Details of circuits and/or installed equipment vulnerable to damage when testing: ....

Details of test instruments used (state serial and/or asset numbers):

Continuity	LM.11
Insulation resistance	LM.14
Earth fault loop impedance	LM.5
RCD	LM.16
Earth electrode resistance	N/A

Tested by:
Name (Capitals): W. WHITE
Signature: W White
Date: 21.7.2015

Circuit number	Circuit Description	Overcurrent device				Conductor details			Ring final circuit continuity (Ω)			Continuity (Ω) (R₁+R₂) or R₂		Insulation Resistance (MΩ)		Polarity	Zₛ (Ω)	RCD			Remarks (continue on a separate sheet if necessary)
		BS (EN)	type	rating (A)	breaking capacity (kA)	Reference Method	Live (mm²)	cpc (mm²)	r₁ (line)	rₙ (neutral)	r₂ (cpc)	(R₁ + R₂)	R₂	Live - Live	Live - Earth	Insert ✓ or X		@ IΔn (ms)	@ 5IΔn (ms)	Test button operation	
1	2	3	4	5	6	7	8	9	10	11	12	13	14	15	16	17	18	19	20	21	22
1	Lights up	3036		5	2	100	1.5	1.0				✓		100	100	✓					Faulty pendants
2	Lights down	"		5	2	100	1.5	1.0				✓		110	110	✓					Faulty pendants
3	Ring up	"		30	2	102	2.5	1.5				✓		120	120	✓					No RCD
4	Ring down	"		30	2	102	2.5	1.5				✓		150	150	✓					No RCD, broken socket
5	Cooker	"		30	2	102	6.0	2.5				✓		120	120	✓					
6	Immersion	"		15	2	102	2.5	1.5				✓		150	160	✓					
7	Shower	"		30	2	102	6.0	2.5				✓		150	140	✓					

Test results

Page 5 of 5

8.4.2 Outcome (observations and recommendations)

The Condition Report form in BS 7671 includes notes for guidance. Within Section K Observations, each observation is required to be allocated a classification code as follows:

- ▶ **C1** Danger present – risk of injury. Immediate remedial action required.
- ▶ **C2** Potentially dangerous – urgent remedial action required.
- ▶ **C3** Improvement recommended.
- ▶ **FI** Further investigation required without delay.

FI (Further Investigation) may also be recorded where the inspection has revealed an apparent deficiency which could not, owing to the extent or limitations of the inspection, be fully identified and further investigation may reveal a code C1 or C2 observation.

The person carrying out the inspection and test is required to provide an overall assessment as to whether the installation is satisfactory or unsatisfactory. It is entirely a matter for the electrically skilled person conducting the inspection to decide on the recommendation code to be assigned to a given observation. The person's own judgement as an electrically skilled person should not be unduly influenced by the client. Remember that the person signing the report is responsible for its content. He/she is required to give advice to the householder as to any repairs necessary to make the installation satisfactory, that is, safe. If the code C1 is allocated to an observation, indicating that it requires urgent attention, then the overall assessment must be that it is unsatisfactory. If the code C2 is allocated to an observation, again, the installation is unsatisfactory.

Examples of categories of observations:

C1 *Immediate remedial action required*

- ▶ Broken equipment where live parts are exposed.
- ▶ Conductive parts live (the result of a fault).
- ▶ Incorrect polarity.

C2 *Potentially dangerous – urgent remedial action required*

- ▶ No means of earthing at origin.
- ▶ RCD installed for fault protection (e.g. in a TT installation) that does not operate when the test button is pressed.
- ▶ No circuit protective conductor in one or more lighting circuits with Class I light fittings (Class I light fittings must be earthed).
- ▶ A borrowed neutral (two circuits using the same neutral).
- ▶ A socket-outlet with no earth connection.

C3 *Improvement recommended*

- ▶ No 30 mA RCD protection to socket circuits.
- ▶ No supplementary bonding in bathroom where it is required.
- ▶ No circuit protective conductor in one or more lighting circuits with Class II light fittings (Class II light fittings do not require to be earthed but may be changed to Class I).
- ▶ Use of water pipe as a means of earthing.

FI   Further investigation

Where further investigation may reveal a code C1 or C2:

▶   a consumer unit is fitted with devices and components of different manufacture and may not meet the requirements of BS EN 61439-3.

▶   it is suspected that devices within a consumer unit are subject to a product recall.

Three-phase supplies

9

This chapter gives a brief introduction to three-phase electricity supplies and the calculation of load current for a.c. equipment.

9.1 Electricity distribution

With particular exceptions in rural areas, electricity is distributed at low voltage by a three-phase distribution system with 230 V a.c. rms nominal voltage to earth and 400 V a.c. rms nominal voltage between phases at a frequency of 50 cycles per second (Hz). Depending on the load of the building, either a single-phase or a three-phase supply will be provided by the electricity distributor.

9.2 Dwellings

Individual dwellings will be supplied at single-phase unless the demand of the premises is high (say, an after-diversity demand over 15 kW), or due to network conditions. Whilst individual flats in a development will almost certainly be supplied at single-phase, the supply brought into the building is likely to be three-phase. A typical PME (protective multiple earthed) installation in a block of flats is shown in Figure 9.2.

▼ **Figure 9.2** Three-phase PME supply to flats

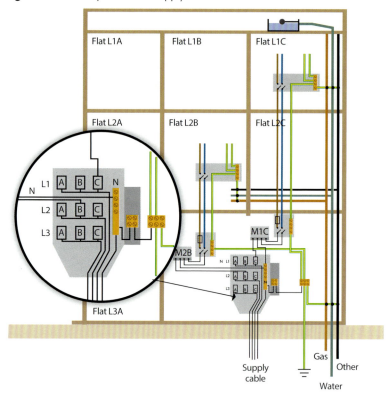

NOTE: For loads generating triple harmonics, as may be found in computer power supplies and electronic controls to lighting, the neutral currents may not sum to zero but would combine. These discussions are outside the scope of this Guide; for further information, refer to the IET publication *Commentary on the IET Wiring Regulations.*

9.3 Characteristics of three-phase supplies

The physical construction of three-phase generators results in the peak voltage of each of the three phases of a three-phase supply being displaced in time by 120 degrees, as shown in Figures 9.3a-c. The result is that the voltage between phases is not 2 × 230 volts as any two phases are never at their peak at the same time, but $\sqrt{3} \times 230$, i.e. 400 volts. The currents in each phase of a symmetrical three-phase load (each phase has the same power factor) are similarly 120 degrees out of phase. The neutral currents of each phase of a three-phase supply return in a common neutral. If the loads are symmetrical and balanced the total neutral current will be zero. This clearly has particular benefits for electricity distribution in that it significantly reduces voltage drop and energy losses.

▼ Figure 9.3a Three-phase supply voltages

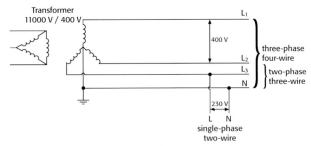

▼ Figure 9.3b Three-phase voltages, with angular displacement

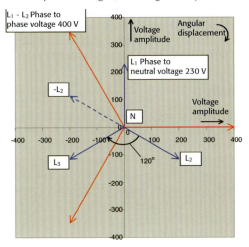

▼ Figure 9.3c Current waveform

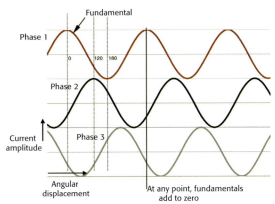

9.4　Fault levels

In the event of a fault, particularly a three-phase fault to earth, the fault current flowing for a three-phase supply will be considerably in excess of that for a single-phase installation. Three-phase equipment will generally require a higher fault rating than single-phase. Fault levels may be 18 kA (0.5 p.f.) at the connection of the service line to the LV distribution main and 25 kA (0.23 p.f.) at the point of connection to the busbars in the distribution substation.

Single-phase earth fault loop impedance testers that also provide a fault current measurement facility are unlikely to give an accurate indication of fault levels, particularly close up to the origin of a three-phase supply. However, for general guidance, fault levels should be estimated as being twice that indicated by a single-phase to earth fault loop impedance tester (a brief explanation would be that the loop impedance tester makes a measurement using single-phase test currents; a three-phase short-circuit fault is likely to result in currents of at least double these as the balanced nature of the fault may result in there being in effect no neutral impedance).

9.5　Design

In the design of a three-phase installation, it is particularly important that there be a clear plan and logic to the electrical installation to facilitate safe working during maintenance. Identifiable locations should be supplied from the same distribution board and particular care taken in labelling. Where equipment in a particular location is supplied from more than one distribution board and more than one phase, there needs to be careful labelling to make clear to electricians carrying out subsequent maintenance that it may be necessary to isolate more than one distribution board or more than one phase of the supply.

NOTE:　With a three-phase supply, while the voltage between phases is 400 volts the voltage to earth remains at 230 volts.

There are particular hazards associated with short-circuits in three-phase supplies; for example, arcs can cause melting and ionising. Care has to be taken when using multi-range test instruments on three-phase supplies, particularly if the instruments do not have fused leads or are not similarly protected.

9.6　Alternating voltage and current

For alternating current supplies, the current in an inductive load, for example, a motor, will lag behind the voltage and in a capacitive circuit the current will lead the voltage, see Figure 9.6a.

▼ **Figure 9.6a**   Voltage and lagging current waveforms

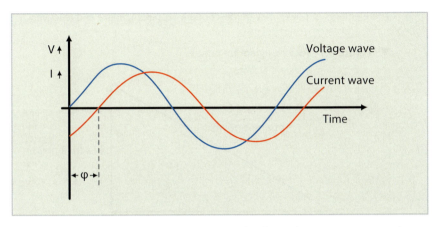

It is very useful to represent these currents and voltages by vectors. A vector has a direction as well as a dimension, see Figure 9.6b.

▼ **Figure 9.6b**   Inductive current and voltage vectors

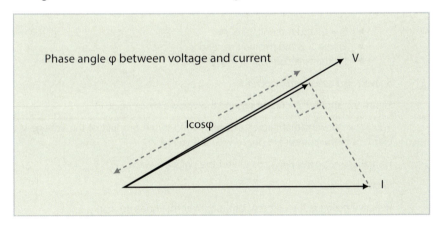

9.7 Complex power

The elements of power may be described as apparent, active and reactive.

▼ **Figure 9.7** Vector diagram for power

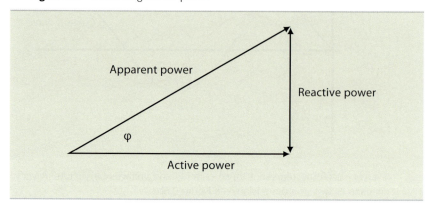

Apparent power S

Apparent power is simply given by:

- ▶ S = VI and the units are VA
- ▶ where V is the rms voltage and
- ▶ I is the rms current.

Active (also called real) power P

Figure 9.7 shows the current lagging the voltage by an angle Ø.

The active power (that carries out real work) is the product of the voltage V and the element of the current I in phase with the voltage V.

Hence active power P=VI cos Ø and the units are watts (W).

Reactive power Q

Reactive power is the element of the power that does no work

Reactive power Q=VI sin Ø and the units are VAr

It is usual in electrical installations to give complex power in kVA and active power P in kW.

9.8   Single-phase a.c. equipment

Apparent power S (in kVA) $= \dfrac{U_0 I_b}{1000}$

Active power P (in kW) $= \dfrac{U_0 I_b \cos \emptyset}{1000}$

hence $I_b = \dfrac{S \text{ (in kVA)} 1000}{U_0}$

and $I_b = \dfrac{P \text{ (in kW)} 1000}{U_0 \cos \emptyset}$

9.9   Three-phase a.c. equipment

Apparent power S (in kVA) $= \dfrac{3 U_0 I_b}{1000}$

Active power P (in kW) $= \dfrac{3 U_0 I_b \cos \emptyset}{1000}$

hence $I_b = \dfrac{S \text{ (in kVA)} 1000}{3 U_0}$

and $I_b = \dfrac{P \text{ (in kW)} 1000}{3 U_0 \cos \emptyset}$

where:

- ▶ P is active power, a measure of ability to carry out work, sometimes called real power.
- ▶ U_0 is the rms voltage to earth.
- ▶ I_b is the rms load current.
- ▶ $\cos \emptyset$ is the power factor of the equipment.

A typical motor rating plate is shown in Figure 9.9.

▼ **Figure 9.9** Motor rating plate

ABB Motors	CE
Type QU 100 L4 AT	IEC 34-1
3 phase Mot GS 2229936 B	cos Ø 0.84
2.2 kW 50 Hz	2.6 kW 60 Hz
1425 r/min	1720 r/min
220-240/380-420 D/Y V	250-280/440-480 D/Y V
8.5 / 4.9	8.8 / 5.1
I.Cl. F	IP 55

Other Building Regulations 10

NOTE: Approved documents and guidance can be freely downloaded from the DCLG website: www.planningportal.gov.uk

Apart from Part P, which is concerned specifically with the safety of fixed electrical installations, there are other requirements in the Building Regulations that affect the electrical installation. Included are:

Approved Document A	–	Structure (depth of chases in walls, and size of holes and notches in floor and roof joists)
Approved Document B	–	Fire safety, Volume 1 – dwellings (fire safety of certain electrical installations; provision of fire alarm and fire detection systems; fire resistance of penetrations through floors and walls)
Approved Document C	–	Site preparation and resistance to contaminants and moisture (moisture resistance of cable penetrations through external walls)
Approved Document E	–	Resistance to the passage of sound (penetrations through floors and walls)
Approved Document F	–	Ventilation (ventilation rates for dwellings)
Approved Document L1	–	Conservation of fuel and power (energy-efficient lighting) L1A new dwellings L1B existing dwellings
Approved Document M	–	Access to and use of buildings (heights of switches and socket-outlets).

10.1 Structure (Approved Document A)

The requirements of Part A are shown below.

Requirement

Loading

A1. (1) The building shall be constructed so that the combined dead, imposed and wind loads are sustained and transmitted by it to the ground

    (a) safely, and

    (b) without causing such deflection or deformation of any part of the building, or such movement of the ground, as will impair the stability of any part of another building.

(2) In assessing whether a building complies with sub paragraph (1) regard shall be had to the imposed and wind loads to which it is likely to be subjected in the ordinary course of its use for the purpose for which it is intended.

Ground Movement

A2. The building shall be constructed so that ground movement caused by

    (a) swelling, shrinkage or freezing of the subsoil, or

    (b) land-slip or subsidence (other than subsidence arising from shrinkage), in so far as the risk can be reasonably foreseen, will not impair the stability of any part of the building.

Disproportionate collapse

A3. The building shall be constructed so that in the event of an accident the building will not suffer collapse to an extent disproportionate to the cause.

The basic requirement for those installing electrical installations in a building is not to cut, drill, chase, penetrate or in any way interfere with the structure so as to cause significant reduction in its load-bearing capability.

10.1.1 Notches and holes in simply supported floor and roof joists

Notches should be no deeper than 0.125 times the depth of a joist and should not be cut closer to the support than 0.07 of the span and not be cut further away from the support than 0.25 times the span.

Holes should be:

▶ no greater diameter than 0.25 times the depth of the joist;

▶ drilled at the neutral axis;

▶ not less than 3 diameters (centre to centre) apart; and

▶ located between 0.25 and 0.4 times the span from the support.

No notches or holes should be cut into the roof rafters, other than at supports where the rafter may be birdsmouthed to a depth not exceeding 0.33 times the rafter depth.

▼ **Table 10.1.1** Joist and stud notch and drill limits

Item	Location	Maximum size
Notching joists	Top edge, 0.07 to 0.25 of span	0.125 × depth of joist
Drilling joists	Centre line, 0.25 to 0.4 of span	0.25 × depth of joist

▼ **Figure 10.1.1** Notches and holes in wooden joists

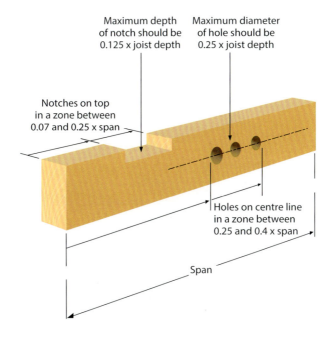

NOTES:

(a) Maximum diameter of hole should be 0.25 × joist depth.
(b) Holes on centre line in a zone between 0.25 and 0.4 × span.
(c) Maximum depth of notch should be 0.125 × joist depth.
(d) Notches on top in a zone between 0.07 and 0.25 × span.
(e) Holes in the same joist should be at least 3 diameters apart.

10.1.2 Chases

(Part A section 2C30.)

Vertical chases should not be deeper than one-third of the wall thickness or in cavity walls one-third of the thickness of the leaf.

Horizontal chases should not be deeper than one-sixth of the thickness of the leaf or wall.

Chases should not be so positioned as to impair the stability of the wall, particularly where hollow blocks are used.

▼ **Figure 10.1.2** Chases in walls

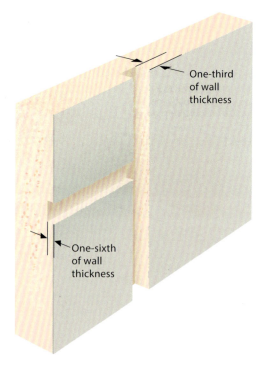

One-third of wall thickness

One-sixth of wall thickness

10.2 Fire safety (Approved Document B)

NOTE: Approved document B1 2006 edition incorporating 2010 and 2013 amendments refers to BS 5839-1:2002 and BS 5839-6:2004. Whilst these are not the latest versions of the standards, the clauses relevant to the text of the approved document are unchanged.

When certifying installations to BS 5839 Parts 1 and 6, installers must follow the requirements and certify to the latest versions. The latest current versions are:

- ▶ BS 5839-1:2013; and
- ▶ BS 5859-6:2013.

See also Figure 10.2.4

10.2.1 The requirement

Requirement B1 of Part B is shown below.

Requirement	Limits on application
Means of warning and escape	
B1. The building shall be designed and constructed so that there are appropriate provisions for the early warning of fire, and appropriate means of escape in case of fire from the building to a place of safety outside the building capable of being safely and effectively used at all material times.	Requirement B1 does not apply to any prison provided under section 33 of the Prisons Act 1952 (power to provide prisons etc.).

Approved Document B states, 'In the Secretary of State's view Requirement B1 will be met if:

- **(a)** there is sufficient means for giving early warning of fire for persons in the building;
- **(b)** there are routes of sufficient number and capacity, which are suitably located to enable persons to escape to a place of safety in the event of fire; and
- **(c)** the routes are sufficiently protected from the effects of fire where necessary.'

10.2.2 Early warning of fire

(a) BS 5839 Fire detection and fire alarm systems

All new dwellinghouses should be provided with a fire detection and fire alarm system in accordance with BS 5839-6 *Code of Practice for the Design and Installation of Fire Detection and Alarm Systems in Dwellings* to at least Grade D Category LD3.

The smoke and heat alarms should comply with BS EN 14604:2005 or BS 5446-2:2003. They should be mains-operated with a standby power supply such as a battery or capacitor.

Specific guidance on meeting this requirement is summarised in section 10.2.2b below. The text from Section 1 of Approved Document B, Volume 1 is found in Appendix C.

NOTE: A Grade D system has one or more mains-powered smoke alarms, each with an integral standby supply. The system may, in addition, incorporate one or more mains-powered heat alarms, each with an integral standby supply. A Category LD3 system is intended for the protection of life and requires detectors in all circulation spaces that form part of the escape routes from the dwelling.

(b) Specific guidance on smoke and heat alarms

Basic requirement

In a standard house (single-storey or multi-storey with no storey exceeding 200 m² floor area), interlinked smoke alarms are to be installed as follows:

(a) In circulation areas between sleeping places and places where fires are most likely to start, e.g. kitchens and living rooms.
(b) In circulation spaces within 7.5 m of the door to each habitable room.
(c) At least one on every storey.
(d) If the kitchen is not separated from the circulation area by a door, a compatible interlinked heat detector or heat alarm must additionally be installed in the kitchen. See Figures 10.2.2a and 10.2.2b.

Positioning of equipment

(e) Ceiling mounted alarms and detectors must be fixed at 300 mm from the walls and light fittings.
(f) The sensors mounted between 25 mm and 600 mm below the ceiling (25 to 150 mm for heat detectors).
(g) Only alarms and detectors suitable for wall mounting are to be wall mounted and are to be fixed above doorway height.
(h) Smoke alarms should not be fixed next to or directly above heaters or air conditioning outlets. They should not be fixed in bathrooms, showers, cooking areas or garages, or any other place where steam, condensation or fumes could give false alarms.
(i) Smoke alarms should not be fitted in places that get very hot (such as a boiler room), or very cold (such as an unheated porch).
(j) All equipment should be safely accessible for routine maintenance including testing and cleaning. They should not be fixed over stairs or openings between floors.

▼ **Figure 10.2.2a** Minimum requirement for smoke alarms (standard house, no storey exceeds 200 m² floor area)

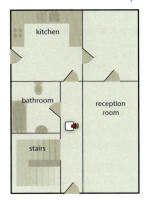
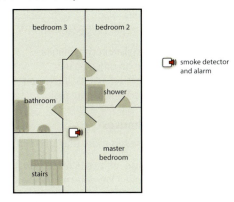

▼ **Figure 10.2.2b** Minimum requirements for smoke alarms (standard house, no storey exceeds 200 m² floor area) with the kitchen not separated from the circulation space by a door

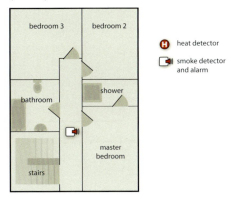

(c) Wiring of smoke and heat alarms

The detectors and alarms are required to:

(a) be linked so that the operation of one will initiate all;

(b) be permanently wired with an independent circuit to the distribution board (consumer unit), or, preferably, supplied from a local, regularly used lighting circuit (there should be a means of isolating the supply to the alarms without switching off the lighting); and

(c) have a standby power supply such as a battery or capacitor.

NOTE: Mains-powered smoke detectors may be interconnected by radio links.

Other than for large houses, the cable for the power supply to each self-contained unit and for the interconnections between self-contained units need have no fire survival properties and needs no special segregation.

Otherwise, fire alarm system cables generally are required to be fire resistant and segregated as per BS 5839-1 and BS 5839-6 to minimise adverse effects from:

- ▶ installation cable faults;
- ▶ fire on other circuits;
- ▶ electromagnetic interference; and
- ▶ mechanical damage.

(d) Large houses

A large house has more than one storey and any one storey of area exceeding 200 m².

A large house of two storeys (excluding basement storeys) should be fitted as a minimum with a fire detection and a Grade B Category LD3 alarm system as described in BS 5839-6:2004.

A large house of three or more storeys (excluding basement storey) should be fitted as a minimum with a Grade A, Category LD2 system as described in BS 5839-6:2004 with detectors sited in accordance with the recommendations of BS 5839-1 for a Category L2 system. See Figure 10.2.2c.

▼ Figure 10.2.2c Outline of fire system for large houses of three or more storeys (Grade A Category LD2 system)

(e) Loft conversions, alterations and extensions

Where additional rooms are provided above ground floor level an automatic smoke detection and alarm system based on linked smoke and heat alarms should be installed to ensure that the occupants of the new rooms are warned of any fire (in the existing or new rooms) that may impede their escape.

Smoke alarms may also be needed where new rooms are provided at ground floor level if a fire within the existing house might impede the escape of the occupants of the new room.

10.2.3 Spread of fire

Part B requires precautions to be taken to inhibit the spread of fire within a building and also requires that the internal linings should adequately resist the spread of flame over their surfaces and have, if ignited, a rate of heat release that is reasonable in the circumstances. Part B sets classifications for lining of walls and ceilings and roof lights, which must be adopted by installers. It is also particularly applicable to thermoplastic materials, which include lighting diffusers forming part of the ceiling.

When carrying out electrical installations in a building the installer must not degrade the precautions taken or building design features intended to limit the spread of fire or limit the propagation of smoke and fumes.

In particular, if compartment walls are penetrated, for example, by cables or trunking, the penetrations may require sealing. Luminaires that penetrate a compartment ceiling must not degrade the fire resistance, see Appendix C.

10.2.4 Installation and commissioning certificate (Grade D systems)

The fire alarm installation is to be designed, installed and tested in accordance with BS 5839-6.

Testing

The wiring of one fire alarm installation may be included in the Electrical Installation Certificate (and schedules) if carried out as part of the electrical installation; otherwise, a minor Electrical Installation Certificate shall be completed (for Grade D systems). In all cases, after successful testing and commissioning, a fire system certificate for the design installation and commissioning shall be completed and given to the person ordering the work, see Figure 10.2.4.

Commissioning

The system is inspected and set to work to confirm that:

(a) all manual call points and fire detectors work, i.e.:
  ▶ smoke detectors are tested to confirm that smoke initiates a fire alarm signal (e.g. by use of test aerosols as recommended by the manufacturer); and
  ▶ heat detectors are tested by a suitable heat source and, if and as recommended by the manufacturer, the heat source should not have the potential to ignite a fire and a live flame should not be used.
(b) fire alarm warning devices (including any provided for deaf or hard of hearing people) work.

Handover

Inspection, test and commissioning certificates (and schedules), manufacturer's instructions and necessary guidance are handed over to the person ordering the work.

▼ **Figure 10.2.4** Fire system certificate (for B, C, D, E and F systems)

BRITISH STANDARD BS 5839-6:2013

Model certificate for Grades B, C, D, E and F

Certificate of design, installation and commissioning* of the fire detection and fire alarm system at:

Address:

.... 10, Some St, ...

.... Anytown, ...

.... PO4 2BS ....

It is certified that the fire detection and fire alarm system at the above address complies with the recommendations of BS 5839-6 for design, installation and commissioning of a Category ...LD3.... Grade .D. system, other than in respect of the following variations:*

.... NONE ....

Brief description of areas protected (only applicable to Category LD2 and PD2 systems).

.... Not applicable ....

The entire system has been tested for satisfactory operation in accordance with the recommendations of 23.3p) of BS 5839-6:2013.*

.... Not applicable ....

Instructions in accordance with the recommendations of Clause 24 of BS 5839-6:2013 have been supplied to:*

.... A Responsible Person ....

Signed: .... A Competent Person ....

Date: .... 1/8/2015 ....

For and on behalf of... Fire system Installer Ltd, ...

.... The Office, Industrial Estate, Somewhere, PD45 7BF ....

* Where design, installation and commissioning are not all the responsibility of a single organization or person, the relevant words should be deleted. The signatory of the certificate should sign only as confirmation that the work for which they have been responsible complies with the relevant recommendations of BS 5839-6:2013. A separate certificate(s) should then be issued for other work.

An Electrical Installation Certificate or a Minor Electrical Installation Works Certificate, as appropriate, has been issued in accordance with BS 7671.

This certificate may be required by an authority responsible for enforcement of fire safety legislation, such as the building control authority or housing authority. The recipient of this certificate might rely on the certificate as evidence of compliance with legislation. Liability could arise on the part of any organization or person that issues a certificate without due care in ensuring its validity.

10.3 Site preparation and resistance to contaminants and moisture (Approved Document C)

The requirements of Part C are shown below.

Persons carrying out electrical work in new buildings must cooperate with the main contractor in complying with the precautions necessary on the site. This will include:

▶ sealing cable entries into the building to prevent the ingress of gas or water; and
▶ taking care that no gas or water seals are penetrated.

Requirement
Preparation of site and resistance to contaminants. **C1.** **(1)** The ground to be covered by the building shall be reasonably free from any material that might damage the building or affect its stability, including vegetable matter, topsoil and pre-existing foundations. **(2)** Reasonable precautions shall be taken to avoid danger to health and safety caused by contaminants on or in the ground covered, or to be covered by the building and any land associated with the building. **(3)** Adequate sub-soil drainage shall be provided if it is needed to avoid: 　**(a)** the passage of ground moisture to the interior of the building; 　**(b)** damage to the building, including damage through the transport of water-borne contaminants to the foundations of the building. **(4)** For the purpose of this requirement, 'contaminant' means any substance which is or may become harmful to persons or buildings including substances which are corrosive, explosive, flammable, radioactive or toxic. Resistance to moisture **C2.** The walls, floors and roof of the building shall adequately protect the building and people who use the building from harmful effects caused by: 　**(a)** ground moisture; 　**(b)** precipitation including wind-driven spray; 　**(c)** interstitial and surface condensation; and 　**(d)** spillage of water from or associated with sanitary fittings or fixed appliances.

10.4 Resistance to the passage of sound (Approved Document E)

The requirements of Part E are shown below.

Electrical installations must not degrade the resistance to sounds of the building. This may require the sealing of cable, conduit and trunking penetrations of walls and ceilings/floors. Trunking and conduit should not be so installed as to conduct sound in contravention of Part E. Unless suitable to address acoustic requirements, luminaires should not breach acoustic barriers. The effects of thermal insulation are discussed in paragraph 10.6.5 of this Guide.

Requirement	*Limits on application*
Protection against sound from other parts of the building and adjoining buildings	
E1. Dwelling-houses, flats and rooms for residential purposes shall be designed and constructed in such a way that they provide reasonable resistance to sound from other parts of the same building and from adjoining buildings.	
Protection against sound within a dwelling-house etc.	
E2. Dwelling-houses, flats and rooms for residential purposes shall be designed and constructed in such a way that (a) internal walls between a bedroom or a room containing a water closet and other rooms, and (b) internal floors provide reasonable resistance to sound.	Requirement E2 does not apply to (a) an internal wall which contains a door, (b) an internal wall that separates an en suite toilet from the associated bedroom, or (c) existing walls and floors in a building which is subject to a material change of use.
Reverberation in the common internal parts of buildings containing flats or rooms for residential purposes	

Requirement	Limits on application
E3. The common internal parts of buildings which contain flats or rooms for residential purposes shall be designed and constructed in such a way as to prevent more reverberation around the common parts than is reasonable.	Requirement E3 only applies to corridors, stairwells, hallways and entrance halls which give access to the flat or room for residential purposes.

Acoustic conditions in schools

E4. (1) Each room or other space in a school building shall be designed and constructed in such a way that it has the acoustic conditions and the insulation against disturbance by noise appropriate to its intended use. (2) For the purposes of this Part, 'school' has the same meaning as in section 4 of the Education Act 1996[4]; 'school building' means any building forming a school or part of a school.	

Particular guidance is given in Approved Document E with respect to socket-outlets that should be applied to all accessories.

In separating walls between rooms and between rooms and corridors:

(a) sockets must be staggered on opposite sides of the separating wall, that is, not back-to-back; and

(b) deep boxes and chases must not be used in separating walls.

Similar guidance is given for sockets on either side of a cavity wall.

For framed walls the guidance is to:

(a) stagger the position of sockets (and other accessories) on opposite sides of the separating wall and to install a layer of the cladding behind the accessory box;

(b) allow a minimum edge-to-edge stagger of 150 mm.

10.5 Ventilation (Approved Document F: 2010 edition)

The requirements of Part F are shown below.

Requirement	Limits on application
Means of ventilation **F1**(1). There shall be adequate means of ventilation provided for people in the building. **F1**(2). Fixed systems for mechanical ventilation and any associated controls must be commissioned by testing and adjusting as necessary to secure that the objective referred to in sub paragraph (1) is met.	Requirement F1 does not apply to a building or space within a building: a. into which people do not normally go; or b. which is used solely for storage; or c. which is a garage used solely in connection with a single dwelling.

Requirements in the Building Regulations 2010
See also regulations 16C, 20AA and 20C requiring the provision of user instructions and, for mechanical systems, testing and commissioning with test results and notice of commissioning being given to the local authority.

10.5.1 Ventilation systems

Approved Document F gives guidance on the provision of ventilation in new dwellings and non-domestic buildings.

The approved document considers four ventilation systems (see Figure 10.5.1):

▶ System 1: Background ventilators and intermittent extract fans
▶ System 2: Passive stack ventilation
▶ System 3: Continuous mechanical extract
▶ System 4: Continuous mechanical supply and extract with heat recovery.

Systems 2 to 4 require specialist knowledge and are more suitable for new buildings.

Electricians will be involved with the installation of system 1 extract fans and systems 3 and 4 whole house mechanical ventilation systems.

System 1 (background ventilators with intermittent extract fans) is the most common for both new and existing dwellings. Typically:

▶ Trickle ventilators (for continuous whole building background ventilation) provide fresh air and deal with background levels of water vapour and pollutants.
▶ Extract fans stop water vapour and pollutants generated intermittently in 'wet rooms' – kitchens, utility rooms (rooms with a sink and laundry appliances), bathrooms (rooms with a bath or shower and possibly a WC) and sanitary accommodation (rooms with a WC) – from spreading to the rest of the building.

▶ Windows and external doors provide purge ventilation to aid the removal of high concentrations of pollutants and water vapour arising occasionally from, for example, painting and decorating, water spillages and burnt food.

▼ **Figure 10.5.1** Ventilation systems

System 1: Background ventilators and intermittent extract fans

System 2: Passive stack ventilation

System 3: Continuous mechanical extract

System 4: Continuous mechanical supply and extract with heat recovery

10.5.2 New dwellings

In a new dwelling, the required 'equivalent' area of trickle ventilators depends on the total floor area and the number of bedrooms. For example, it varies from 25,000 mm² for a 1-bedroom dwelling with a floor area up to 50 m², to 105,000 mm² for a 5-bedroom dwelling with a floor area of 200 m².

Table 10.5 gives the minimum intermittent extract ventilation fan rates for kitchens, utility rooms, bathrooms and sanitary accommodation. Fan manufacturers' guidance should identify which rooms are suitable for which fans.

The approved document also gives ventilation rates for continuous extract as an alternative to intermittent extract.

For purge ventilation, the overall area of any windows and external doors in a habitable room (a room other than a wet room) should be at least 1/20 of the floor area of the room. If a window opens less than 30 degrees and there are no external doors, its area should be 1/10 of the floor area. If the habitable room has no external walls, purge ventilation can be provided through an adjacent room or conservatory.

▼ **Table 10.5.1** System 1 extract ventilation rates for new dwellings

Room	Minimum intermittent extract ventilation fan rate
Kitchen	30 litres/second adjacent to hob* 60 litres/second elsewhere
Utility room	30 litres/second
Bathroom (including shower room)	15 litres/second
Sanitary accommodation (separate from bathroom)	6 litres/second or purge ventilation

* e.g. extract cooker hood

In wet rooms, purge ventilation can be provided by an openable window of any size. If there are no external walls, extract ventilation will suffice (although it will take longer to purge the room), but the extract fan should have a 15-minute overrun.

For further guidance on System 1 design and installation, ventilation of basements, passive stack ventilation, and the whole house mechanical ventilation Systems 3 and 4, refer to Approved Document F.

10.5.3 Existing dwellings

Section 7 of Approved Document F covers work on existing buildings. Generally, when doing new building work in an existing building, the new work must comply with all relevant parts of the Building Regulations, and compliance of the rest of the building with the Regulations must be made no worse.

Wet rooms

Section 7 gives specific guidance when adding a wet room – either building a new wet room as or part of an extension, or converting an existing dry room (e.g. a bedroom or part of a bedroom) into a wet room.

The ventilation requirements when adding a wet room can be met by fitting an intermittent extract fan along with a background ventilator of at least 2,500 mm^2 equivalent area. There should also be a 10 mm gap (or equivalent) under the door. If the wet room has no external walls to fit a background ventilator or a window for purge ventilation, the fan should have a 15-minute overrun.

Alternatively, ventilation can be provided by:

▶ a continuously running single room heat recovery ventilator; or
▶ a 'passive stack' ventilator; or
▶ a continuous extract fan.

Refurbishing kitchens and bathrooms

When carrying out building work, including electrical installation work, compliance with other requirements of the Building Regulations, including ventilation in an existing kitchen or bathroom, must not be made worse than before the work commenced. If there is an existing extract fan (including a cooker hood extracting to outside) or passive stack ventilator you should retain or replace it. If there is no existing ventilation system you need not provide one.

10.6 Conservation of fuel and power in dwellings (Approved Documents L1A and L1B: 2010 editions)

The requirement of Part L of Schedule 1 to the Building Regulations 2010 is shown below:

Requirement

Part L Conservation of fuel and power
L1. Reasonable provision shall be made for the conservation of fuel and power in buildings by:
 	
(a) limiting heat gains and losses –
    (i) through thermal elements and other parts of the building fabric; and
    (ii) from pipes, ducts and vessels used for space heating, space cooling and hot water services;
(b) providing fixed building services which –
    (i) are energy efficient;
    (ii) have effective controls; and
    (iii) are commissioned by testing and adjusting as ~ necessary to ensure they use no more fuel and power than is reasonable in the circumstances; and
(c) providing to the owner sufficient information about the building, the fixed building services and their maintenance requirements so that the building can be operated in such a manner as to use no more fuel than is reasonable in the circumstances.

Building Regulations 2010 See also regulations 7 materials and workmanship, 23 thermal elements, 26C target CO_2 emission rate, 28 consequential improvements and 29 energy performance certificates.

Approved Documents L1A and L1B and the 'second tier' document *Domestic Heating Compliance Guide* give guidance on meeting the Part L requirement when installing the following electrical systems in new and existing dwellings:

- ▶ lighting;
- ▶ electric heating systems, including electric boilers, warm air systems, panel heaters and storage heaters;
- ▶ underfloor heating;
- ▶ heat pumps; and
- ▶ individual domestic (micro) combined heat and power (microCHP).

The installation of solar photovoltaic panels is currently covered by a DTI *Guide to the installation of PV systems*.

This Guide considers electric lighting and electric heating only.

10.6.1 Fixed internal lighting

In new buildings and for extensions and rewires the fixed internal and external lighting should meet the recommended minimum standards for efficacy and controls as below.

(a) in the areas affected by the building work, provide low-energy light fittings (fixed lights or lighting units) that number not less than three-quarters (3/4) of all the light fittings in the main dwelling spaces of those areas (excluding infrequently accessed spaces used for storage such as cupboards and wardrobes).

(b) low-energy light fittings should have lamps with a luminous efficacy greater than 45 lamp lumens per circuit-watt and a total output greater than 400 lamp lumens.

(c) light fittings whose supplied power is less than 5 circuit-watts are excluded from the overall count of the total number of light fittings.

Light fittings may be either:

(i) dedicated fittings which will have separate controlgear and will take only low-energy lamps (e.g. pin-based fluorescent or compact fluorescent lamps); or

(ii) standard fittings supplied with low-energy lamps with integrated controlgear, e.g. bayonet or Edison screw-based compact fluorescent lamps.

Light fittings with GLS tungsten filament lamps or tungsten halogen lamps would not meet the standard.

The Energy Saving Trust publication GIL 20 *Lower-energy domestic lighting* gives guidance on identifying suitable locations for fixed energy-efficient lighting.

10.6.2 Fixed external lighting

Where fixed external lighting is installed, light fittings must be provided with:

(a) either:
(i) lamp capacity not greater than 100 lamp-watts per light fitting;
(ii) all lamps automatically controlled so as to switch off after the area lit by the fitting becomes unoccupied; and
(iii) all lamps automatically controlled so as to switch off when daylight is sufficient.

(b) or
(i) lamp efficacy greater than 45 lumens per circuit-watt;
(ii) all lamps automatically controlled so as to switch off when daylight is sufficient; and
(iii) light fittings controllable manually by occupants.

NOTE: The guidance in 10.6.1 and 10.6.2 is taken from HM Government publication *Domestic Building Services Compliance Guide* available as a free download.

10.6.3 Electric boilers serving wet central heating systems

NOTES:

Specific guidance is provided by the DCLG in:

- ▶ *Domestic Building Services Compliance Guide* and
- ▶ *Domestic heating compliance guide, compliance with Approved Documents L1A: new dwellings and L1B: existing dwellings.*

These are available as free downloads.

The minimum requirements when installing a new system in a new or existing dwelling are:

- ▶ Circuits for heating and hot water should be fully pumped.
- ▶ Boilers should be fitted with flow temperature control and be capable of modulating power input to the primary water.
- ▶ If the boiler also supplies domestic hot water (DHW), the boiler should have a boiler control interlock so that the boiler and pump are switched off when there is no call for heat. The use of thermostatic radiator valves (TRVs) alone is not sufficient.
- ▶ Dwellings with a floor area up to 150 m² should be divided into at least two space heating zones with independent temperature control, one of which is assigned to the living area.
- ▶ Dwellings with a floor area greater than 150 m² should be provided with at least two space heating zones, each having separate timing and temperature controls.
- ▶ Sub-zoning is not appropriate in single-storey open-plan dwellings in which the living area is greater than 70 per cent of the total floor area.
- ▶ A room thermostat or programmable room thermostat should be installed in the main zone to control temperature; in each of the other zones, a room thermostat or programmable room thermostat should be installed, or alternatively an individual radiator control such as a TRV on each radiator.
- ▶ Time control of space and water heating should be provided by a full programmer with separate timing to each circuit; or separate timers for each circuit; or a programmable room thermostat to the heating circuit with separate timing of the hot water circuit.
- ▶ Unless the appliances comply as a whole with the appropriate British Standard, protection against overheating should be provided by means of an appropriate non-self-resetting device functioning independently of the thermostat.
- ▶ Hot-water cylinders should have provision for two thermostatically controlled electric heating elements or immersion heaters, the lower one connected to use off-peak electricity and the upper for boost operation.

The minimum requirements when installing a complete replacement system are the same as for a new system.

The minimum requirements when replacing a component are:

- When replacing a boiler, an existing system with semi-gravity circulation should be converted to fully pumped.
- When replacing a boiler, new controls, as for new systems, should be installed unless they are already installed and fully operational.
- If an individual component of the control system is being replaced, for example, a room thermostat, it is not necessary to upgrade the system to meet the minimum provisions.

10.6.4 Electric heating other than boilers
NOTES:

Specific guidance is provided by the DCLG in:

- *Domestic Building Services Compliance Guide* and
- *Domestic heating compliance guide, compliance with Approved Documents L1A: new dwellings and L1B: existing dwellings.*

These are available as free downloads.

For electric warm air heating systems, the minimum requirements are:

- Time and temperature control by either a time switch/programmer and room stat; or a programmable room thermostat. The controls may be integral to the heater or external.
- Zoning as for central heating boiler systems – at least two space heating zones with independent temperature control for floor areas up to 150 m², and at least two zones with separate timing and temperature controls for floor areas greater than 150 m².
- Protection against overheating should be provided such that the heating elements cannot be activated until the prescribed air flow has been established and are deactivated when the air flow is less than the prescribed value.

For electric panel heaters, the minimum requirement is local time and temperature control, provided by integral or external time switches and thermostats.

For electric storage heaters, the minimum requirements are:

- automatic control of input charge based on measurement of internal temperature;
- temperature control by adjustment of heat release using a damper or other thermostatically controlled means; and
- the frame and enclosure of space heating appliances should be of non-combustible material.

10.6.5 Thermal insulation

Care must be taken with the design and installation of electrical installations in dwellings with high levels of thermal insulation. Circuits designed for cable installation methods 100, 101 and 102 (see 2.3.1 and 4.1) must not be totally enclosed in thermal insulation but must be in contact with the ceiling/wall or joist. The builder must be made aware of the electrical installation thermal design limitations, that is, the need to keep cables in contact with a thermally conductive surface on one side. Luminaires must not be covered with thermal insulation unless suitable.

10.7 Access to and use of buildings (Approved Document M)

The requirements of Part M are shown below.

Requirement	Limits on application
Access and Use **M1.** Reasonable provision shall be made for people to **(a)** gain access to, and **(b)** use the building and its facilities.	The requirements of this Part do not apply to **(a)** man extension of or material alteration of a dwelling, or **(b)** many part of a building which is used solely to enable the building or any service or fitting in the building to be inspected, repaired or maintained.
Access to Extensions to Buildings and other Dwellings **M2.** Suitable independent access shall be provided to the extension where reasonably practicable.	Requirement M2 does not apply where suitable access to the extension is provided through the building that is extended.
Sanitary Conveniences in Extensions to Buildings other than Dwellings **M3.** If sanitary conveniences are provided in any building that is to be extended, reasonable provision shall be made within the extension for sanitary conveniences.	Requirement M3 does not apply where there is reasonable provision for sanitary conveniences elsewhere in the building, such that people occupied in, or otherwise having occasion to enter the extension, can gain access to and use those sanitary conveniences.

Requirement	Limits on application
Sanitary Conveniences in Dwellings **M4. (1)** Reasonable provision shall be made in the entrance storey for sanitary conveniences, or where the entrance storey contains no habitable rooms, reasonable provision for sanitary conveniences shall be made in either the entrance storey or principal storey. **(2)** In this paragraph 'entrance storey' means storey which contains the principal entrance and 'principal storey' means the storey nearest the entrance storey which contains a habitable room, or if there are two such storeys equally near, either such storey.	

Part M of Schedule 1 includes in its requirements that reasonable provision must be made for disabled people to gain access to and use a building.

10.7.1 Heights of switches and socket-outlets

The Building Regulations require switches and socket-outlets in dwellings to be installed so that all persons, including those whose reach is limited, can easily use them. A way of satisfying the requirement is to install switches, socket-outlets and controls throughout the dwelling in accessible positions and at a height of between 450 mm and 1200 mm from the finished floor level (see Figure 10.7.1).

Because of the sensitivity of circuit-breakers, RCCBs and RCBOs fitted in consumer units/fuseboards should be readily accessible.

The guidance given in Approved Document M applies to all new dwellings. Note that if a dwelling is rewired, there is no requirement to provide the measures described above providing that upon completion the building is no worse in terms of the level of compliance with the other Parts of Schedule 1 to the Building Regulations.

▼ **Figure 10.7.1** Height of switches, socket-outlets, etc.

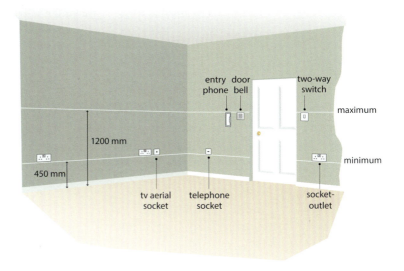

10.7.2 Consumer units in new dwellings

Approved Document M does not recommend a height for new consumer units. However, Part P advises that 'one way of complying with Part M in new dwellings is to mount consumer units so that the switches are between 1350 mm and 1450 mm above floor level. At this height, the consumer unit is out of reach of young children yet accessible to other people when standing or sitting'.

Identification of conductors · 11

11.1 Introduction

The requirements of BS 7671 were harmonized with the technical intent of CENELEC Standard HD 384.5.514: Identification, including 514.3: Identification of conductors, now withdrawn.

The cable standards have been harmonized with CENELEC Harmonization Document HD 308 S2: 2001 *Identification of cores in cables and flexible cords*. These standards specify the cable core marking, including cable core colours, to be implemented in the CENELEC countries.

This chapter provides guidance on marking at the interface between old and harmonized colours, and general guidance on the colours to be used for conductors.

British Standards for fixed and flexible cables have been harmonized with the colours in HD 308 S2. BS 7671 has been modified to align with these cable colours, but also allows other suitable methods of marking connections by colour (tapes, sleeves or discs) or by alphanumerics (letters and/or numbers). Methods may be mixed within an installation.

▼ **Table 11.1** Identification of conductors

Function	Alphanumeric	Colour
Protective conductors		Green-and-Yellow
Functional earthing conductor		Cream
a.c. power circuit[1]		
Line of single-phase circuit	L	Brown
Neutral of single- or three-phase circuit	N	Blue
Line 1 of three-phase a.c. circuit	L1	Brown
Line 2 of three-phase a.c. circuit	L2	Black
Line 3 of three-phase a.c. circuit	L3	Grey
Two-wire unearthed d.c. power circuit		
Positive of two-wire circuit	L+	Brown
Negative of two-wire circuit	L−	Grey
Two-wire earthed d.c. power circuit		
Positive (of negative earthed) circuit	L+	Brown
Negative (of negative earthed) circuit[2]	M	Blue
Positive (of positive earthed) circuit[2]	M	Blue
Negative (of positive earthed) circuit	L−	Grey
Three-wire d.c. power circuit		
Outer positive of two-wire circuit derived from three-wire system	L+	Brown
Outer negative of two-wire circuit derived from three-wire system	L−	Grey
Positive of three-wire circuit	L+	Brown
Mid-wire of three-wire circuit[2,3]	M	Blue
Negative of three-wire circuit	L−	Grey
Control circuits, ELV and other applications		
Line conductor	L	Brown, Black, Red, Orange, Yellow, Violet, Grey, White, Pink or Turquoise
Neutral or mid-wire[4]	N or M	Blue

NOTES:
1 Power circuits include lighting circuits.
2 M identifies either the mid-wire of a three-wire d.c. circuit, or the earthed conductor of a two-wire earthed d.c. circuit.
3 Only the middle wire of three-wire circuits may be earthed.
4 An earthed PELV conductor is blue.

11.2 Addition or alteration to an existing installation

11.2.1 Single-phase

An addition or an alteration made to a single-phase installation need not be marked at the interface provided that:

(a) the old cables are correctly identified by the colours red for line and black for neutral; and

(b) the new cables are correctly identified by the colours brown for line and blue for neutral.

A warning label must be provided at the consumer unit or distribution board.

▼ **Figure 11.2.1** Addition to a single-phase installation

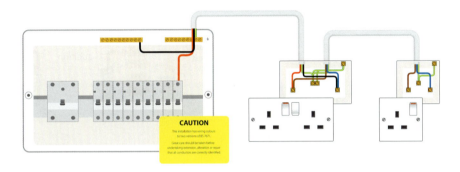

11.2.2 Two- or three-phase installation

Where an addition or an alteration is made to a two- or a three-phase installation wired in the old core colours with cable to the new core colours, unambiguous identification is required at the interface. Cores should be marked as follows:

Neutral conductors
Old and new conductors: N

Line conductors
Old and new conductors: L1, L2, L3.

▼ **Table 11.2.2** Example of conductor marking at the interface for additions and alterations to an a.c. installation identified with the old cable colours

Function	Old conductor		New conductor	
	Colour	Marking	Marking	Colour
Line 1 of a.c.	Red	L1	L1	Brown*
Line 2 of a.c.	Yellow	L2	L2	Black*
Line 3 of a.c.	Blue	L3	L3	Grey*
Neutral of a.c.	Black	N	N	Blue
Protective conductor	Green-and-Yellow			Green-and-Yellow

* Three single-core cables with insulation of the same colour may be used if identified at the terminations.

▼ **Figure 11.2.2** Addition to a three-phase installation

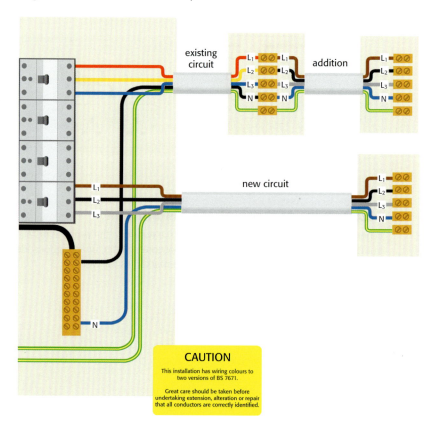

CAUTION

This installation has wiring colours to two versions of BS 7671.

Great care should be taken before undertaking extension, alteration or repair that all conductors are correctly identified.

11.3 Switch wires in a new installation or an addition or alteration to an existing installation

Where a two-core cable with cores coloured brown and blue is used as switch wires, both conductors being line conductors, the blue conductor must be marked brown or L at its terminations.

▼ **Figure 11.3**   One-way switch

NOTE: The sheaths of cables entering the ceiling rose have been cut back for clarity in order to show the core colours. Sheaths must enclose the cable cores except within accessories.

11.4 Intermediate and two-way switch wires in a new installation or an addition or alteration to an existing installation

Where a three-core cable with cores coloured brown, black and grey is used as switch wires, all three conductors being line conductors, the black and grey conductors must be marked brown or L at their terminations.

▼ **Figure 11.4**   Two-way switching

NOTE: The sheaths of cables entering the ceiling rose have been cut back for clarity in order to show the core colours. Sheaths must enclose the cable cores except within accessories.

11.5 Line conductors in a new installation or an addition or alteration to an existing installation

Line conductors should be coloured as shown in Table 11.1.

In a two- or three-phase power circuit, the line conductors may all be of one of the permitted colours and either identified L1, L2, L3 or marked brown, black, grey at their terminations to identify the phases.

11.6 Changes to cable core colour identification

▼ **Table 11.6a** Cable to BS 6004 (flat cable with bare cpc)

Cable type	Old core colours	New core colours
Single-core + bare cpc	Red or Black	Brown or Blue
Two-core + bare cpc	Red, Black	Brown, Blue
Alt. two-core + bare cpc	Red, Red	Brown, Brown
Three-core + bare cpc	Red, Yellow, Blue	Brown, Black, Grey

▼ **Table 11.6b** Standard 600/1000 V armoured cable to BS 6346, BS 5467 or BS 6724

Cable type	Old core colours	New core colours
Single-core	Red or Black	Brown or Blue
Two-core	Red, Black	Brown, Blue
Three-core	Red, Yellow, Blue	Brown, Black, Grey
Four-core	Red, Yellow, Blue, Black	Brown, Black, Grey, Blue
Five-core	Red, Yellow, Blue, Black, Green-and-Yellow	Brown, Black, Grey, Blue, Green-and-Yellow

▼ **Table 11.6c** Flexible cable to BS 6500

Cable type	Old core colours	New core colours
Two-core	Brown, Blue	No change
Three-core	Brown, Blue, Green-and-Yellow	No change
Four-core	Black, Blue, Brown, Green-and-Yellow	Brown, Black, Grey, Green-and-Yellow
Five-core	Black, Blue, Brown, Black, Green-and-Yellow	Brown, Black, Grey, Blue, Green-and-Yellow

11.7   Addition or alteration to a d.c. installation

When an addition or an alteration is made to a d.c. installation wired in the old core colours with cable to the new core colours, unambiguous identification is required at the interface.

Cores must be marked as follows:

Neutral and midpoint conductors
Old and new conductors: M

Line conductors
Old and new conductors: Brown or Grey, or
Old and new conductors: L, L+ or L−

▼ **Table 11.7**   Example of conductor marking at the interface for additions and alterations to a d.c. installation identified with the old cable colours

Function	Old conductor colour	Old conductor marking	New conductor marking	New conductor colour
Two-wire unearthed d.c. power circuit				
Positive of two-wire circuit	Red	L+	L+	Brown
Negative of two-wire circuit	Black	L−	L−	Grey
Two-wire earthed d.c. power circuit				
Positive (of negative earthed) circuit	Red	L+	L+	Brown
Negative (of negative earthed) circuit	Black	M	M	Blue
Positive (of positive earthed) circuit	Black	M	M	Blue
Negative (of positive earthed) circuit	Blue	L−	L−	Grey
Three-wire d.c. power circuit				
Outer positive of two-wire circuit derived from three-wire system	Red	L+	L+	Brown
Outer negative of two-wire circuit derived from three-wire system	Red	L−	L−	Grey
Positive of three-wire circuit	Red	L+	L+	Brown
Mid-wire of three-wire circuit	Black	M	M	Blue
Negative of three-wire circuit	Blue	L−	L−	Grey

Scottish Building Reg 12

Additional information relevant to building regulations in Scotland.

12.1 Introduction

This chapter explains the requirements for electrical installations in Scotland as covered by the Building (Scotland) Act 2003 and associated legislation. Detailed information on the Scottish system including building regulations can be found at the Scottish Government Building Standards Division (BSD) website: www.scotland.gov.uk/bsd.

Requirements for electrical installations in Scotland are addressed by standard 4.5 – electrical safety for all buildings, standard 4.6 – electrical fixtures for domestic buildings only and standard 4.8 – danger from accidents for all buildings (see 12.5.1, 12.5.2 and 12.5.3) and, as in England and Wales, persons carrying out electrical installations must ensure that the work they carry out both complies with building regulations and the relevant functional standards.

There are no significant differences in general installation requirements for electrical work, with Scotland, England and Wales citing BS 7671 (as amended) as the recommended means of satisfying building standards requirements. However, Part P electrical self-certification schemes in England and Wales do not apply to work in Scotland. In Scotland, qualified and experienced electricians can certify that their installation work meets the requirements of building regulations under the Certification of Construction (Electrical Installations to BS 7671) scheme approved under section 7(2) of the Building (Scotland) Act 2003.

12.2 Legislation

12.2.1 Building (Scotland) Act 2003

This Act gives Scottish Ministers the power to make building regulations to secure the health, safety, welfare and convenience of persons in or about buildings and of others who may be affected by buildings or matters connected with buildings; to further the conservation of fuel and power; and to further the achievement of sustainable development. The Act also allows Ministers to issue guidance documents in support of these regulations.

12.2.2 The Building (Scotland) Regulations 2004 (as amended)

The Building (Scotland) Regulations 2004 (as amended) are made under the powers of the Building (Scotland) Act 2003 and apply solely in Scotland.

The regulations apply to building work such as alterations, extensions, conversions, erections and demolition of buildings and also to the provision of services, fittings and equipment in, or in connection with, buildings except where they are specifically exempted from regulations 8 to 12 (schedule 1 to regulation 3). Schedule 3 to regulation 5 lists work that must comply but does not need a building warrant.

The regulations prescribe functional standards for buildings, which can be found in schedule 5 to regulation 9. The regulations and functional standards are amended periodically. However, it is the regulations in force at the time of the application that must be complied with.

12.2.3 Responsibility for compliance with building regulations

In Scotland, final responsibility for compliance with building regulations rests with the 'relevant person' (defined in section 17 of the Act), who will normally be the owner or developer of a building. However, any person carrying out work, including electrical work, has a duty to ensure their work complies with building regulations.

12.3 Scottish guidance documents

12.3.1 Technical guidance

The BSD Technical Handbooks are published in two volumes, domestic buildings and non-domestic buildings. Each handbook is split into seven sections. Section 0 covers general issues and offers an introduction and guidance to the Building Regulations. Sections 1 to 7 include the functional standards together with guidance on how to comply with them:

- ▶ Section 1: Structure
- ▶ Section 2: Fire
- ▶ Section 3: Environment
- ▶ Section 4: Safety
- ▶ Section 5: Noise
- ▶ Section 6: Energy
- ▶ Section 7: Sustainability

The Technical Handbooks (current and previous) may be found on the BSD website at: www.scotland.gov.uk/bsd. Installers working in Scotland should familiarise themselves with the Scottish system through the information available on the BSD website and from the certification guidance published by the scheme providers, SELECT and NICEIC, who both operate a certification scheme (see12.4.3) for electrical installations.

12.3.2 Procedural guidance

The BSD have published a Procedural Handbook that explains the Scottish system in more detail and is accessible on the BSD website. Unlike the Technical Handbooks, this handbook has no specific legal status, but is designed to clarify the procedures.

12.3.3 Certification guidance

The BSD have published a Certification Handbook for schemes under section 7(2) of the Building (Scotland) Act 2003, which is accessible on the BSD website. The handbook provides guidance on the optional procedure using Approved Certifiers of Construction (see 12.4.3).

12.4 Procedures

In Scotland, a building warrant is required for building work unless:

- ▶ it is not a 'building' as defined in the Building (Scotland) Act 2003;
- ▶ it is exempt from building regulations (schedule 1 to regulation 3); and
- ▶ it does not require a warrant but must still comply with building regulations (schedule 3 to regulation 5).

Anyone intending to carry out electrical work that requires a warrant should note that:

- ▶ a building warrant must be obtained before work starts;
- ▶ a completion certificate must be submitted when work is complete; and
- ▶ a completion certificate must be accepted by the verifier before occupation or use of a new building (or extension) is permitted.

12.4.1 Work requiring a building warrant

Except where described above a building warrant must be applied for and be granted by the verifier before work can commence. It is an offence to commence work without a warrant. The application is made to the local authority building standards department (currently only the 32 local authorities have been appointed as verifiers for their geographical area), who will assess the application and issue the building warrant if proposals are considered to comply with building regulations.

The use of an Approved Certifier of Design or Construction does not remove the need to obtain a building warrant but the certificate they issue must be accepted by the verifier. If an Approved Certifier of Design or Construction is used the warrant fee can be discounted.

NOTE: A discount will only be available for a Certificate of Construction when the verifier is notified on the application for building warrant that it is intended that an Approved Certifier of Construction will be used.

The verifier must be notified when work commences. Notification is usually made by the relevant person (normally the owner or developer of the building) or his agent.

Designers and installers are directed to section 2 – Fire of the Scottish Building Standards Technical Handbooks for detailed guidance on issues including openings and fire stopping (standards 2.1, 2.2 and 2.4), escape route lighting (standard 2.10), fire alarm and detection systems (standard 2.11) and automatic life safety fire suppression systems (standard 2.15). A comparison between Scottish and English guidance documents is included in paragraph 12.7.

12.4.2 Completion certificate

If a building warrant has been granted, a Completion Certificate must be submitted to the verifier once works are complete. This states that works have been completed in accordance with building regulations and the building warrant. If the verifier, after making their reasonable inquiries, is satisfied, they must accept the Completion Certificate. If it is rejected, the verifier will identify the reason for rejection.

Where an Approved Certifier of Construction for electrical installations is used, they will issue a Certificate of Construction which the verifier must accept as proof of compliance for the specific work described. It should be submitted with the Completion Certificate.

12.4.3 Approved Certifiers of Design/Approved Certifiers of Construction

The Building (Scotland) Act 2003 establishes a role for suitably qualified people, businesses or other bodies, when appointed by the Scottish Ministers, to certify that certain design or construction work complies with the Building Regulations. Two roles are designated, Approved Certifiers of Design and Approved Certifiers of Construction, both of which certify compliance with the Building Regulations, as laid down in the scope of the certification scheme run by the scheme provider. Further information on certification schemes may be found on the BSD website.

Approved Certifiers of Construction are responsible for the construction or installation of specified parts of a building, such as the electrical installation, and must have due regard for compliance with the full range of building standards, not just those applicable to the part of the building covered.

The approved schemes for the Scottish Building Regulations are found on the BSD website: http://www.certificationregister.co.uk

There are three Certification of Design schemes. These are Certification of Design (Building Structures) provided by SER Ltd; Certification of Design (Section 6 – Energy) in Domestic Buildings provided by BRE and RIAS; and Certification of Design (Section 6 – Energy) in Non-Domestic Buildings provided by BRE.

There are two Certification of Construction schemes. These are the Certification of Construction (Electrical Installations to BS 7671) scheme provided by SELECT and NICEIC and the Certification of Construction (Drainage, Heating and Plumbing) scheme provided by SNIPEF.

Details of the electrical installation scheme may be obtained from the Scottish Building Services Certification portal at www.sbsc.uk.net

A person who is not a member of such a scheme may still carry out electrical installations and separate guidance is offered to assist verifiers in determining compliance of electrical work.

The guidance for verifiers is available at www.scotland.gov.uk/bsd

12.4.4 Buildings and services exempt from building regulations

Certain buildings and services are exempt from regulations 8 to 12 and a building warrant is not required. They are set out in schedule 1 to regulation 3 of the Building (Scotland) Regulations 2004 (as amended) and include exceptions where the building regulations do apply. The exempt types include:

- ▶ buildings or work covered by other legislation;
- ▶ buildings or work not frequented by people;
- ▶ specialised buildings or work where application of the regulations is largely inappropriate;
- ▶ buildings or works that are minor where enforcement is not in the public interest; and
- ▶ temporary buildings or works.

12.4.5 Works not requiring building warrant

Certain works subject to building regulations do not require a building warrant but works must still meet the regulations. They are set out in schedule 3 to regulation 5 of the Building (Scotland) Regulations 2004 (as amended) and include exceptions where a warrant is required.

For example:

- ▶ Any work to or in a house with a storey height not exceeding 4.5 m but only if the exceptions do not apply (type 1).
- ▶ Any work to or in a non-residential building that members of the public do not have access to, with a storey height not exceeding 7.5 m, but only if the exceptions do not apply (type 2).

In both cases the exceptions include, but are not limited to, structural work, work to a separating wall and alterations to roofs or external walls.

There are other specific types of work that can be done to any building that also do not require a warrant (types 3 to 26).

Two matrices identifying electrical work not requiring a warrant have been prepared jointly by the Local Authority Building Standards Scotland (LABSS) and the BSD. They cover domestic and non-domestic work and are accessible on the BSD website (see 12.6).

12.5 Electrical installations

12.5.1 Standard 4.5 – electrical safety

Electrical installations must comply with standard 4.5 (electrical safety), which states that every building must be designed and constructed in such a way that the electrical installation does not threaten the health and safety of the people in and around the building, and does not become a source of fire.

Guidance to standard 4.5 cites BS 7671 as a means of complying with the functional standard and addresses, in brief, general low voltage installations, extra-low voltage installations and installations operating above low voltage.

12.5.2 Standard 4.6 – electrical fixtures

Electrical installations must comply with standard 4.6 (electrical fixtures), which states that every domestic building must be designed and constructed in such a way that electric lighting points and socket-outlets are provided.

Guidance to standard 4.6 addresses lighting, light switches in common areas, entryphone systems and socket-outlets.

12.5.3 Standard 4.8 – danger from accidents

Electrical installations must comply with standard 4.8 (danger from accidents), which states that every building must be designed and constructed in such a way that manual controls for ventilation and for electrical fixtures can be operated safely.

Guidance to standard 4.8 includes that electrical sockets, switches and other controls should be sited to allow safe and convenient use.

12.5.4 Fitness and durability of materials, workmanship and access for maintenance

Regulation 8 requires that materials, fittings and components used should be suitable for their purpose, correctly used or applied, and sufficiently durable, taking account of normal maintenance practices, to meet the requirements of the regulations. It also implements the intention of the Construction Products Regulation, that specification of construction products should not be used to effectively bar the use of construction products or processes from other European countries. The relevant countries are those in the European Union, and those who in the European Economic Area Act of 1993 agreed to adopt the same standards.

12.6 Guidance on electrical work not requiring a warrant

12.6.1 Domestic buildings

▼ **Building (Scotland) Regulations 2004 Regulation 5, Schedule 3**
Guidance on electrical work not requiring a warrant

The Scottish
Government
Riaghaltas na h-Alba

DOMESTIC BUILDINGS		WORK TO EXISTING BUILDINGS		
	Type[1]	Flat	House (up to 2 storeys)	House (3 storeys & above)
Repairs and replacement				
Rewiring[2]	24	required	not required	required
Electrical fixtures, e.g. luminaires	24	not required	not required	not required
New work				
Electrical work affected by demolition or alteration of the roof, external walls or elements of structure	1	required	required	required
Electrical work adversely affecting a separating wall, e.g. recessed sockets	1	required	required	required
New power socket outlets	1	required	not required	required
Mains operated fire alarm system	1	required	not required	required
Electrical work to automatic opening ventilators (including auto-detection)	1	required	not required	required
Electrically operated locks	1	required	not required	required
Wiring to artificial lighting	1	required	not required	required
Wiring to emergency lighting	1	required	not required	required
Electrical work associated with sprinkler system	1	required	not required	required
Electrical work associated with new boiler (large)	1	required	not required	required
Electrical work associated with new boiler (small)	6	not required	not required	not required
Electrical work associated with new shower	11, 12	not required	not required	not required
Electrical work associated with new extract fan	13	not required	not required	not required
Extra-low voltage installations	22	not required	not required	not required

NOTES:

1   Building work type as referenced in schedule 3.
2   A building warrant is not required for rewiring where it is a repair or replacement works to a level equal to the installation (or part thereof) being repaired or replaced.

12.6.2 Non-domestic buildings

▼ **Building (Scotland) Regulations 2004 Regulation 5, Schedule 3**

Guidance on electrical work not requiring a warrant

NON-DOMESTIC BUILDINGS	WORK TO EXISTING BUILDINGS		
	Non-residential buildings with a storey, or creating a storey, not more than 7.5 m		Other non-domestic buildings
Type[1]	**No public access**[2]	**Public access**	
Repairs and replacement			
Rewiring[3]	24	not required	required	required
New work			
Electrical work affected by demolition or alteration of the roof, external walls or elements of structure	2	required	required	required
Electrical work adversely affecting a separating wall, e.g. recessed sockets	2	required	required	required
Electrical work adversely affecting a loadbearing wall	2	required	required	required
New power socket-outlets	2	not required	required	required
Automatic fire detection system	2	not required	required	required
Electrical work to automatic opening ventilators	2	not required	required	required
Electrical work to automatic fire dampers	2	not required	required	required
Electrically operated locks	2	not required	required	required
Wiring to artificial lighting	2	not required	required	required
Wiring to emergency lighting	2	not required	required	required
Outdoor luminous tube signs[4]	2	not required	not required	not required
Electrical work associated with new boiler (large)	2	not required	required	required

NON-DOMESTIC BUILDINGS	WORK TO EXISTING BUILDINGS			
	Non-residential buildings with a storey, or creating a storey, not more than 7.5 m			Other non-domestic buildings
	Type[1]	No public access[2]	Public access	
Electrical work associated with new boiler (small)	6	not required	not required	not required
Electrical work associated with new shower	11,12	not required	not required	not required
Electrical work associated with new extract fan	13	not required	not required	not required
Extra-low voltage installations	22	not required	not required	not required

NOTES:

1 Building work type as referenced in schedule 3.

2 Non-residential buildings to which the public does not have access may include:
   ▶ Existing offices
   ▶ Existing storage buildings
   ▶ Existing industrial buildings, e.g. factories and workshops
   ▶ Existing assembly and entertainment buildings not open to the public, e.g. some educational buildings and private members clubs.
   Non-residential buildings to which the public has access may include:
   ▶ Existing assembly and entertainment buildings open to the public, e.g. community schools, pubs and clubs.

3 A building warrant is not required for rewiring where it is a repair or replacement works to a level equal to the installation (or part thereof) being repaired or replaced.

4 Subject to the Town and Country Planning (Control of Advertisement) (Scotland) Regulations 1984.

12.7 Comparison between Scottish and English guidance documents

The following table lists the Approved Documents and their equivalents within the Scottish Building Standards system, the Technical Handbooks.

Scotland	England and Wales
Section 1 (structure)	AD A (Structure)
Section 2 (fire)	AD B (Fire safety)
Section 3 (environment) – standards 3.1 to 3.4 and 3.10 & 3.15	AD C Site preparation and resistance to contaminants and moisture)
No equivalent	AD D (Toxic substances)
Section 5 (noise)	AD E (Resistance to the passage of sound)
Section 3 (environment) – standards 3.14 (ventilation) and 3.10 (precipitation)	AD F (Ventilation)
Section 3 (environment) – standard 3.12 and section 4 (safety) – standard 4.9 (danger from heat)	AD G (Hygiene)
Section 3 (environment) – standards 3.5 to 3.9 (drainage) and 3.25 & 26 (waste storage)	AD H (Drainage and waste disposal)
Section 3 (environment) – standards 3.17 to 3.24 and section 4 (safety) – standard 4.11	AD J (Combustion appliances and fuel storage systems)
Section 4 (safety) – standards 4.3, 4.4, 4.8 & 4.12	AD K (Protection from falling, collision and impact)
Section 6 – (energy)	AD L (Conservation of fuel and power)
Section 3 (environment) – standards 3.11 & 3.12 and section 4 (safety) – standards 4.1, 4.2, 4.3, 4.6, 4.7 & 4.10	AD M (Access to and use of buildings)
Section 4 (Safety) – standard 4.8	AD N (Glazing – safety in relation to impact, opening and cleaning)
Section 4 (Safety) – standard 4.5	AD P (Electrical safety – Dwellings)
Section 7 (Sustainability)	No equivalent

In addition, the Technical Handbooks contain functional standards for which there are no direct equivalents in England and Wales. These are 'facilities in dwellings', 'heating', 'natural lighting' and 'security' (standards 3.11, 3.13, 3.16 and 4.13).

Whilst the application of the principles outlined within chapter 10 is equally valid in Scotland, detailed recommendations within the BSD guidance may differ. Designers and installers are therefore advised to familiarise themselves with the standards and guidance within the Scottish Building Standards Technical Handbooks prior to undertaking work in Scotland.

Welsh Building Regulations 13

13.1 Introduction

On 31 December 2011 the power to make building regulations for Wales was transferred to Welsh Ministers. This means Welsh Ministers will make any new building regulations or publish any new building regulations guidance applicable in Wales from that date.

The Building Regulations 2010 and related guidance, including approved documents as at that date, will continue to apply in Wales until Welsh Ministers make changes to them. For example, the 2006 version of Approved Document P applies at the date of publication of the Guide, see section 13.2.

As guidance is reviewed and changes made Welsh Ministers will publish separate approved documents.

A Welsh Part L (Conservation of fuel and power) came into effect in July 2014. A Welsh Part B (Fire safety) requires automatic fire suppression in new and converted residential buildings, which will be extended to new and converted houses and flats etc. in January 2016.

13.2 Approved Document P: Electrical safety – Dwellings: 2006 edition with 2010 amendments

Reproduced here is the guidance given in the previous edition of this Guide (section 1.3).

▼ **Figure 13.2** The requirement of Part P of the Building Regulations

Requirement	Limits on application
Design and installation	
P1. Reasonable provision shall be made in the design and installation of electrical installations in order to protect persons operating, maintaining or altering the installations from fire or injury.	The requirements of this part apply only to electrical installations that are intended to operate at low or extra-low voltage and are – (a) in or attached to a dwelling; (b) in the common parts of a building serving one or more dwellings, but excluding power supplies to lifts; (c) in a building that receives its electricity from a source located within or shared with a dwelling; or (d) in a garden or in or on land associated with a building where the electricity is from a source located within or shared with a dwelling.

13.2.1 Scope of Part P

Part P applies to electrical installations in buildings or parts of buildings comprising:

(a) dwellinghouses and flats;
(b) dwellings and business premises that have a common metered supply – for example shops and public houses with a flat above with a common meter;
(c) common access areas in blocks of flats such as corridors and staircases; and
(d) shared amenities of blocks of flats such as laundries and gymnasiums.

Part P applies also to parts of the above electrical installations:

(e) in or on land associated with the buildings – for example, Part P applies to fixed lighting and pond pumps in gardens; and
(f) in outbuildings such as sheds, detached garages and greenhouses.

Part P does not apply to:

(a) business premises in the same building as dwellings with separate metering; and

(b) lifts in blocks of flats.

▼ **Figure 13.2.1** Scope of Part P

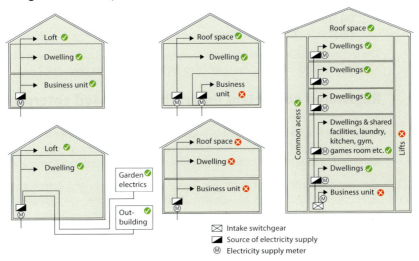

13.2.2 Compliance with Part P

In the Secretary of State's view, the requirements of Part P will be met by adherence to the 'Fundamental Principles' for achieving safety given in BS 7671:2008 Chapter 13.

To achieve these requirements electrical installations in dwellings, etc. must be:

▶ designed and installed to afford appropriate protection against mechanical and thermal damage, and so that they do not present electric shock and fire hazards to people; and

▶ suitably inspected and tested to verify that they meet the relevant equipment and installation standards.

Chapter 13 of BS 7671 is met by complying with Parts 3 to 7 of BS 7671 except as allowed by regulations 120.3 and 120.4. That is, any intended departure from these Parts requires special consideration by the designer of the installation and is to be noted on the Electrical Installation Certificate specified in Part 6 (Regulation 120.3). Where the use of a new material or invention leads to departures from the Regulations, the resulting degree of safety of the installation is to be not less than that obtained by compliance with the Regulations. Such use is to be noted on the Electrical Installation Certificate specified in Part 6.

This publication is written to provide simple rules and installation requirements including circuit specifications for compliance.

13.3 Notification to Building Control

Except as below, the relevant building control body must be notified of all proposals to carry out electrical installation work in dwellings, etc. (see section 1.3) before the work begins.

It is not necessary to give prior notification of proposals to carry out electrical installation work in dwellings to building control bodies if the work:

(a) is carried out by a registered competent enterprise; or

(b) is non-notifiable minor work.

The technical requirements of Part P apply to all electrical installation work in dwellings, whether the work needs to be notified to Building Control or not.

Registered competent enterprise (Assessed enterprise)

It is not necessary to give prior notification of proposals to carry out electrical installation work in dwellings to building control bodies if a competent enterprise registered with an electrical self-certification scheme authorised by the Secretary of State undertakes the work. Such a registered competent enterprise may be called an assessed enterprise.

The registered competent enterprise is responsible for ensuring compliance with BS 7671:2008+A3:2015 and all relevant Building Regulations. On completion of the work, the registered competent enterprise is required to formally declare that the work complies with parts 4 and 7 of the Building Regulations and notify its self-certification scheme in accordance with its procedures who, after checking competence to carry out the work, will notify Building Control (see Figure 13.3). The person ordering the work should also receive a duly completed Electrical Installation Certificate 2008 as per, or similar to, the models in BS 7671:2008+A3:2015. As required by BS 7671, the certificate must be made out and signed on behalf of the enterprise by the competent person or persons that carried out the design, construction, inspection and testing work. Copies of relevant BS 7671:2008+A3:2015 model forms are shown in Chapter 6 and guidance is given on inspection, testing and the completion of forms.

Persons working on behalf of a registered competent enterprise are not required to notify Building Control in advance of work. The registered competent enterprise will notify its registration body on completion of the work and the registration body will then notify Building Control.

Persons employed by a registered competent enterprise but working on their own behalf and not registered individually with a self-certification scheme are required to notify Building Control of any work within the scope of the Building Regulations they intend to carry out.

▶ **Figure 13.3** Building Regulations notification

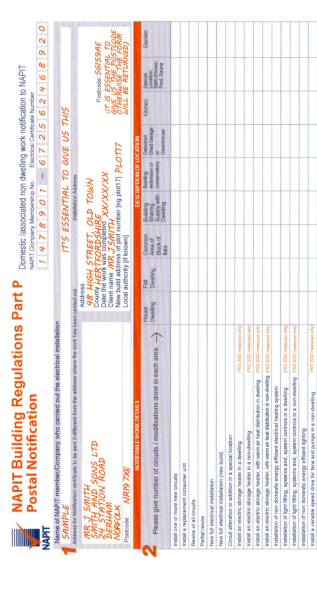

Unregistered competent persons

Competent persons not registered with an assessed enterprise are required to notify Building Control of notifiable work before work starts (emergency work is to be notified as soon as possible). Inspection and testing should be carried out as per BS 7671. Completed certificates with schedules should be forwarded to Building Control who will take the certificates into account when deciding what further action needs to be taken.

On notification, Building Control become responsible for the work. If Building Control decide the completed work is safe and meets the requirements of all the Building Regulations, they will on request issue a Building Control certificate.

Members of bodies such as the Institution of Engineering and Technology (IET) who carry out electrical work in their own homes are not exempt from the requirement to notify Building Control, in the same way that members of the Institution of Civil Engineers who do work on their house foundations are not exempt.

Unqualified installers

Installers (contractors or DIYers) not qualified to inspect and test their work must notify Building Control of notifiable work before the work starts (emergency work is to be notified as soon as possible). Building control is then responsible for ensuring that the work is safe, including arranging for inspection and testing as necessary.

Compliance certificates

On completion (including inspection and testing) of notifiable work, the householder is to receive:

- ▶ a Building Regulations compliance certificate (issued by the self-certification scheme on behalf of the registered competent enterprise); or
- ▶ a building control completion certificate issued by the building control of the local authority; or
- ▶ a final certificate issued by approved inspectors; and
- ▶ an appropriate Electrical Installation Certificate complete with schedules.

13.4   Non-notifiable minor work

Table 1 of Part P: Work that need not be notified to Building Control bodies

Work consisting of:
- ▶ Replacing any fixed equipment (for example, socket-outlets, control switches and ceiling roses) which does not include the provision of any new fixed cabling
- ▶ Replacing the cable for a single circuit only, where damaged, for example, by fire, rodent or impact[a]
- ▶ Re-fixing or replacing the enclosures of existing installation components[b]
- ▶ Providing mechanical protection to existing fixed installations[c]
- ▶ Installing or upgrading main or supplementary equipotential bonding[d]

Work that is not in a kitchen or special location[e] and does not involve a special installation[d] and consists of:
- ▶ Adding lighting points (light fittings and switches) to an existing circuit[f]
- ▶ Adding socket-outlets and fused spurs to an existing ring or radial circuit[f]

Work not in a special location on:
- ▶ Telephone or extra-low voltage wiring and equipment for the purpose of communications, information technology, signalling, control and similar purposes
- ▶ Prefabricated equipment sets and associated flexible leads with integral plug and socket connections

NOTES:

a   On condition that the replacement cable has the same current-carrying capacity and follows the same route.
b   If the circuit's protective measures are unaffected.
c   If the circuit's protective measures and current-carrying capacity of conductors are unaffected by increased thermal insulation.
d   Such work shall comply with other applicable legislation, such as the Gas Safety (Installation and Use) Regulations
e   Special locations and installations are listed in Table 2.
f   Only if the existing circuit protective device is suitable and provides protection for the modified circuit, and other relevant safety provisions are satisfactory.

Table 2 of Part P: Special locations and installations[a]

Special locations:
- ▶ Locations containing a bath tub or shower basin
- ▶ Swimming pools or paddling pools
- ▶ Hot air saunas

> **Table 2 of Part P: Special locations and installations[a]**
>
> **Special installations:**
> - ▶ Electric floor or ceiling heating systems
> - ▶ Garden lighting or power installations
> - ▶ Solar photovoltaic (PV) power supply systems
> - ▶ Small-scale generators such as microCHP units
> - ▶ Extra-low voltage lighting installations, other than pre-assembled, CE-marked lighting sets
>
> **NOTE:** See IET Guidance Note 7, which gives guidance on special locations.

It is not necessary to give prior notification of proposals to carry out the minor electrical installation work described in Table 1 of Approved Document P and where it does not include the provision of a new circuit.

However, non-notifiable work is required to comply with the technical requirements of Part P, including BS 7671 requirements for inspection and testing and the issuing of a minor works certificate.

BS 7671 certificates can usually only be issued by the person who carried out the electrical work. The guidance in Approved Document P, however, does allow DIYers undertaking non-notifiable work to have their work checked by a competent third party.

Additional notes from Approved Document P

Tables 1 and 2 above give the general rules for determining whether or not electrical installation work is notifiable. The rules are based on the risk of fire and injury and what is practicable. The following notes provide additional guidance and specific examples.

Notifiable

a   Notifiable jobs include new circuits back to the consumer unit, extensions to circuits in kitchens and special locations (bathrooms, etc.) and locations associated with special installations (garden lighting and power installations, etc.).

Not notifiable

b   Replacement (like for like), repair and maintenance jobs are generally not notifiable, even if carried out in a kitchen, special location or location associated with a special installation.

Consumer unit

c   Consumer unit replacements are, however, notifiable.

Bathrooms

d   In large bathrooms, the location containing a bath or shower is defined by the walls of the bathroom.

Conservatories and attached garages

e Conservatories and attached garages are not special locations. Work in them is therefore not notifiable unless it involves the installation of a new circuit or the extension of a circuit in a kitchen or special location or associated with a special installation.

Detached garages and sheds

f Detached garages and sheds are not special locations. Work within them is notifiable only if it involves new outdoor wiring or a new circuit.

Outdoor lighting

g Outdoor lighting and power installations are special installations. Any new work in, for example, the garden or work that involves crossing the garden is notifiable.

Fixed equipment

h The installation of fixed equipment is within the scope of Part P even where the final connection is by a 13 A plug and socket. However, work is notifiable only if it involves fixed wiring and the installation of a new circuit or the extension of a circuit in a kitchen or special location or associated with a special installation.

Outside lights, sockets and equipment

i The installation of equipment attached to the outside wall of a house (for example, security lighting, air conditioning equipment and radon fans) is not notifiable under Part P provided that there are no exposed outdoor connections and the work does not involve the installation of a new circuit or the extension of a circuit in a kitchen or special location or associated with a special installation.

j The installation of a socket-outlet on an external wall is notifiable, since the socket-outlet is an outdoor connector that could be connected to cables that cross the garden and require RCD protection.

Modular systems

k The installation of prefabricated, 'modular' systems (for example, kitchen lighting systems and cable systems designed for garden installations) linked by plug and socket connectors is not notifiable, provided that products are CE-marked and that any final connections in kitchens and special locations are made to existing connection units or points (possibly a 13 A socket-outlet).

Gates and garage doors

l Work to connect an electric gate or garage door to an existing isolator is not notifiable, but installation of the circuit up to the isolator is notifiable.

Cookers and electric showers

m The fitting and replacement of cookers and electric showers is not notifiable unless a new circuit is needed.

Central heating control wiring

n New central heating control wiring installations are notifiable even where work in kitchens and bathrooms is avoided.

13.5 Provision of information

After carrying out work, including a new installation, sufficient information shall be provided to the person ordering the work for passing on to the occupant so that persons wishing to operate, maintain or alter an electrical installation can do so with reasonable safety.

Meeting the requirements of BS 7671 will require the installer to:

(a) provide a schedule of inspections, schedule of tests, and an electrical installation certificate or periodic inspection (condition) report (or minor works certificate if appropriate) – see chapters 6 and 8;

(b) provide labelling of the installation as in section 3.4;

(c) install cables in the building fabric only as permitted in section 2.3; and

(d) provide operating instructions and logbooks and, for large or complex installations, detailed plans.

The Electrotechnical Assessment Specification

The Electrotechnical Assessment Specification (EAS) describes:

- ▶ the minimum requirements for an enterprise (e.g. contractor) to be recognised by a certification body as competent to undertake electrical installation work (design, construction, installation and verification) in England and Wales. It includes the minimum technical competence requirements for enterprises to be considered competent to carry out electrical installation work in dwellings in accordance with Part P of the Building Regulations.
- ▶ the competence requirements for registered Qualified Supervisors and Responsible Persons of the competent enterprises.
- ▶ particular requirements for compliance with the Scottish Building Standards.
- ▶ interpretation of the general requirements for bodies operating product certification (including process and service) schemes of BS EN 45011.

The EAS has been prepared by a management committee that includes representatives of the competent person scheme providers, trade associations, the Department for Communities and Local Government, the Electrical Safety Council and the Institution of Engineering and Technology (IET). The IET has accepted ownership of the specification, and provides administrative support to the management committee.

The specification was prepared, as part of the IET support of the electrical industry, (and other interested parties) in introducing electrical safety into the Building Regulations.

The Electrotechnical Assessment Specification is published by the IET, and is available for download in PDF format.

http://electrical.theiet.org/building-regulations

The EAS included in this book has been reproduced with those areas of text that have changed since the last version marked by a line against the changed text.

A

EAS 15-350

ELECTROTECHNICAL *ASSESSMENT* SPECIFICATION FOR USE BY *CERTIFICATION* AND *REGISTRATION BODIES*

February 2015

NOTE: This EAS (February 2015) replaces the previous EAS (October 2012).

Principally a new appendix has been added to modify certain criteria of the scheme document to allow for the entertainment and events industry.

Any changes to the EAS (October 2012) have been indicated by the use of side bars.

1

ELECTROTECHNICAL *ASSESSMENT* SPECIFICATION

A) INTRODUCTION

1. PURPOSE

1.1 This Specification is intended for use by *Certification* and *Registration Bodies* undertaking the *Assessment* of *Enterprise*s carrying out *Electrical installation work*.

1.2 The Electrotechnical *Assessment* Specification (hereafter referred to as the "Specification") has been drawn up by bodies representing electrical installation industry and consumer safety interests in order to enhance the standard of safety of *Electrical installation work*.

1.3 The Specification sets out the minimum requirements to be met by an *Enterprise* in order to be recognized by a *Certification* or *Registration Body* as technically competent to undertake the design, construction, maintenance, verification and/or inspection and testing of one or more of the work categories listed in Appendix 1. *Certification* and *Registration Bodies* will need to develop their own *Scheme* requirements around the minimum *Criteria* set out in this Specification.

2. ACKNOWLEDGEMENT

2.1 Organisations that contributed to the production of this Specification are:

Ascertiva Group Ltd (trading as NICEIC)
Benchmark Certification Ltd
British Standards Institution (BSI)
Certsure LLP (trading as NICEIC and ELECSA)
City & Guilds
Department for Communities and Local Government (DCLG)
Electrical Contractors' Association (ECA)
Electrical Contractors' Association of Scotland (SELECT)
Electrical Safety First
Excellence, Achievement & Learning Limited (EAL)
Fire Industry Association (FIA)
Health and Safety Executive (HSE)
Institution of Engineering and Technology (IET)
Joint Industry Board for the electrical contracting industry (JIB)
NAPIT Registration Ltd
National Association of Professional Inspectors and Testers (NAPIT)
Oil Firing Technical Association (OFTEC)
Society of Electrical and Mechanical Engineers Serving Local Government (SCEME)
SummitSkills
United Kingdom Accreditation Service (UKAS)
Welsh Government

2.2 Organisations that contributed to the production of Appendix 9 are:

British Broadcasting Corporation (BBC)
Independent Television (ITV)
National Outdoor Events Association (NOEA)
Professional Lighting and Sound Association (PLASA)
British Sky Broadcasting (BSKYB)
Association of Event Venues (AEV)
Exhibition Suppliers and Services Association (ESSA)
Broadcasting, Entertainment, Cinematographic & Theatre Union (BECTU)
Association of Show and Agricultural Organisations (ASAO)
Local Authorities Event Organisers Group (LAEOG)
The Event Services Association (TESA)
Production Services Association (PSA)
Association of British Theatre Technicians (ABTT)
Film and TV Services (FTVS)
Ambassador Theatre Group (ATG)
Earls Court & Olympia (EC&O)
DK Services Group
1st Option Safety Services

2

A

3. TABLE OF CONTENTS

[*1] Please note that BS EN 45011 whilst still current has been replaced by ISO/IEC 17065. There is a 3 year transition for accredited *Certification Bodies* to comply with the new standard. However, a new *Certification Body* could seek accreditation to BS EN 45011 and then comply with the transition deadline which is Sept 14th 2015.

3

4.   SCOPE

4.1   This Specification sets out the minimum requirements to be met by *Enterprises* in order to be recognized by *Certification* and *Registration Bodies* as technically competent to undertake *Electrical installation work*.

4.2   The categories of *Electrical installation work* to which this Specification relates are listed in Appendix 1.

4.3   This Specification includes requirements relating to the resources, facilities, personnel and technical standard of *Electrical installation work* of the *Enterprise* being assessed. The *Criteria* for *Assessment* and the requirements for reporting the outcome are also included.

4.4   Particular requirements for *Competent Person Scheme Operator*s are set out in Appendix 6. These requirements include those previously set out in the Minimum Technical Competence document. Particular requirements for Scottish Building Standards are set out in Appendix 7.

5.   DEFINITIONS

Assessed Enterprise – an *Enterprise* which has been assessed in accordance with this Specification as competent in one or more of the categories of work listed in Appendix 1, and which possesses a current *Assessment Certificate*.

Assessment – objective examination of an *Enterprise* including its technical reference documents, test instruments, insurance, samples of *Electrical installation work* and completed certificates and reports, in order to determine the technical competence of the *Enterprise* to carry out *Electrical installation work* in accordance with the relevant reference documents listed in Appendix 2.

Assessment Certificate – a certificate awarded by a *Certification or Registration Body* to an *Assessed Enterprise* (see the *Criteria* at Appendix 5).

Accredited body – a *Certification Body* accredited by UKAS.

Awarding organisation – an organisation approved by OFQUAL, CCEA or SQA to offer regulated *Qualifications.*

Certification Body – an organisation which undertakes the *Assessment* of the technical competence of *Enterprises* in accordance with this Specification and which is accredited against the requirements of BS EN 45011[*1] by UKAS or an equivalent European or international body.

Certificate of competence – a certificate awarded by a *Certification Body* accredited to ISO 17024

Competent Person – a person, considered by the *Enterprise* to possess the necessary technical knowledge, skill and experience to undertake assigned *Electrical installation work*, and to prevent danger and where appropriate injury.

Competent Person Scheme Operator - see Appendix 6

Contracting Office – a location from which an *Enterprise* manages *Electrical installation work*.

Criteria – standards, laws or rules by which a correct judgement can be made by a *Certification* or *Registration Body.*

4

Enterprise – a business undertaking *Electrical installation work* in one or more of the categories listed in Appendix 1. The business may be a sole trader, partnership, limited liability company, public limited company, public authority or other organisation carrying out *Electrical installation work*.

Electrical installation work – the design, construction, maintenance, verification and/or inspection and testing of one or more of the work categories listed in Appendix 1.

Functionality – The ability of a completed installation to operate as required by the relevant standard(s) listed in Appendix 2.

National Occupational Standards (NOS) – Standards that reflect the competence requirements of a defined job role and/or occupation.

Non-conformity – the absence of, or a failure to implement and maintain, one or more required management system elements or a situation which would, on the basis of available objective evidence, raise significant doubt as to the technical standard of the *Electrical installation work* an *Enterprise* is carrying out.

Principal Duty Holder – the person appointed by an *Enterprise* to have responsibility for the maintenance of the overall standard and safety of *Electrical installation work.*

Proposed Qualified Supervisor – a person who meets the requirements of Appendix 4 but has yet to successfully complete an *Assessment* as determined by a *Certification* or *Registration Body.*

Qualification – an OFQUAL, CCEA or SQA regulated award that is based on approved *National Occupational Standards* and delivered through an *Awarding organisation.*

Qualified Supervisor – a *Competent Person* with specific responsibility on a day to day basis for the safety, technical standard and quality of *Electrical installation work.*

Registered Qualified Supervisor – a *Qualified Supervisor* who has been assessed and accepted by a *Certification* or *Registration Body.*

Registration Body – an organisation which undertakes the *Assessment* of the technical competence of *Enterprises* in accordance with this Specification and which is not accredited against the requirements of BS EN 45011[*1] by UKAS or an equivalent European or international body.

Responsible Person – see Appendix 6.

Registered Responsible Person – see Appendix 6.

Scheme – a systematic arrangement of *Criteria.*

6.   WORK SUB-LET BY AN *ASSESSED ENTERPRISE*

6.1   If an *Assessed Enterprise* sub-lets work in any category listed in Appendix 1, the work that is sub-let shall be required to be either:

   a) undertaken by an *Assessed Enterprise* that has a current *Assessment Certificate* covering that particular work category; or

   b) certified as compliant with the relevant standard(s) by an *Assessed Enterprise* that has a current *Assessment Certificate* covering that particular work category.

5

B) REQUIREMENTS RELATING TO THE ASSESSMENT OF THE TECHNICAL COMPETENCE OF ENTERPRISES

7. *ELECTRICAL INSTALLATION WORK*

7.1   The *Enterprise* shall be required to be directly engaged in carrying out *Electrical installation work* in one or more of the categories listed in Appendix 1.

8. TECHNICAL REFERENCE DOCUMENTS

8.1   The *Enterprise* shall be required to have current editions, including all amendments, of Statutory Regulations and of technical reference documents appropriate to the range, scale and category(s) of work for which an *Assessment Certificate* is being sought or has been granted. Appendix 2 details the particular documents to be held for each category of work.

9. TEST INSTRUMENTS

9.1   The *Enterprise* shall be required to have an adequate number of serviceable test instruments and test leads appropriate to the range, scale, geographical spread and category(s) of *Electrical installation work* undertaken.

9.2   The *Enterprise* shall be required to have records demonstrating the accuracy and consistency of test instruments held or hired for the certification of, or for reporting on the condition of, *Electrical installation work*. Guidance on this requirement is given at Appendix 3.

9.3   Where test instruments are hired, the *Enterprise* shall be required to provide evidence of the hire arrangement, together with confirmation of their calibration.

10. CERTIFICATION AND REPORTING

10.1   The *Enterprise* shall be required to issue appropriate certificates and reports in accordance with the relevant standards for all *Electrical installation work* carried out. For *Assessment* purposes, the *Enterprise* shall, at any time, be required to have available copies of all certificates and inspection reports issued during at least the preceding three years.

10.2   If an *Assessed Enterprise* carries out *Electrical installation work* in categories in which it does not have a current *Assessment Certificate*, such work shall be required to be certified by an *Assessed Enterprise* that has a current *Assessment Certificate* for that particular work category.

6

11. PERSONNEL

11.1   The *Enterprise* shall be required to employ persons to carry out *Electrical installation work* who are competent and/or adequately supervised to ensure safety during and on completion of the work.

11.2   The *Enterprise* shall be required to appoint a *Principal Duty Holder* and, for each *Contracting Office*, nominate at least one *Qualified Supervisor* as appropriate to the range, scale, geographical spread and categories of the *Electrical installation work* undertaken from that *Contracting Office*. A *Principal Duty Holder* shall be required to be a principal or employee of the *Enterprise*. A *Principal Duty Holder* may also be a *Qualified Supervisor*. A person nominated as a *Qualified Supervisor* for the purpose of complying with this Specification shall be required to be a *Competent Person* and shall be subject to acceptance by the *Certification* or *Registration Body*.

Responsibilities of the *Principal Duty Holder*

11.3   A *Principal Duty Holder* shall be required to be responsible for ensuring that the *Enterprise* carries out work in accordance with the relevant standards, including the issue of appropriate certificates or inspection reports, as defined in clause 10.1, for all *Electrical installation work* carried out.

11.4   A *Principal Duty Holder* shall be required to be responsible for ensuring that the *Enterprise* undertakes the work activity in compliance with all relevant statutory requirements.

11.5   A *Principal Duty Holder* shall be required to have an understanding of and, for the purposes of this Specification, required to be responsible for, the health and safety and other statutory requirements relating to the *Electrical installation work* being undertaken by the *Enterprise*.

11.6   A *Principal Duty Holder* shall be required to ensure that all *Electrical installation work* is assigned to the *Enterprise*'s *Qualified Supervisor*(s).

11.7   Where a *Registered Qualified Supervisor* ceases to be employed in that capacity, the *Principal Duty Holder* shall be required to immediately notify the *Certification* or *Registration Body*.

11.8   A *Principal Duty Holder* shall be required to ensure that electrical personnel receive any necessary training.

Requirements relating to a *Registered Qualified Supervisor*

11.9   A *Registered Qualified Supervisor* shall be required to have direct responsibility, on a day to day basis, for the safety, quality and technical standard of the *Electrical installation work* carried out by the *Enterprise*. A *Qualified Supervisor* shall be required to ensure that the results of the verification process are accurately recorded on the appropriate certificates or inspection reports.

11.10 A *Registered Qualified Supervisor* shall be required to have adequate knowledge, experience and understanding of the design, construction, maintenance, verification and/or inspection and testing procedures for *Electrical installation work* in accordance with the relevant competence requirements in Appendices 4, 6 and 9.

11.11 *Registered Qualified Supervisor*s shall be required to hold an appropriate BS 7671 *Qualification* awarded by an *Accredited body* within two years of a change to the BS 7671 regulations coming into effect or be able to demonstrate an equivalent level of knowledge.

7

12. INSURANCE

12.1 The *Enterprise* shall be required to hold at least £2 million of public liability insurance covering all work being assessed or within the scope of its *Assessment Certificate*

13. APPLICATION FOR CERTIFICATION OR REGISTRATION

13.1 An *Enterprise* seeking certification or registration under this Specification shall be required to make a written application to a *Certification* or *Registration Body*, stating the *Contracting Office*(s) from which electrical work is managed and the categories of work for each *Contracting Office* for which certification or registration is sought.

13.2 The application, signed by the *Principal Duty Holder* on behalf of the *Enterprise*, is to be required to include the trading title and address of the *Contracting Office* of the *Enterprise*.

13.3 Full details of the *Principal Duty Holder* and of each *Proposed Qualified Supervisor* shall be required to be submitted with the application.

13.4 The *Enterprise* shall be required to undergo technical *Assessment* in accordance with the requirements of Section 15.

13.5 Where an *Enterprise* holds, has held, or has had cancelled, a previous *Assessment Certificate*, the *Enterprise* shall be required to declare this to the *Certification* or *Registration Body* on application for *Assessment*.

13.6 An *Assessed Enterprise* wishing to extend the scope of its certification or registration shall be required to make a written application in accordance with this Section.

14. SURVEILLANCE

14.1 In order to give assurance that the *Assessed Enterprise* is continuing to comply with the requirements of this Specification, the *Assessed Enterprise* shall be subjected to surveillance visits, normally at annual intervals, during the course of which it shall undergo technical *Assessment*.

14.2 A *Certification* or *Registration Body* may, following risk *Assessment* and at its discretion, vary the interval between surveillance visits. The interval shall not exceed three years. The surveillance programme shall identify the planned intervals.

14.3 A surveillance visit shall be required when a *Registered Qualified Supervisor* ceases to be employed in that capacity and the suitability of the proposed replacement is to be assessed in accordance with the procedures defined by the *Certification* or *Registration Body*.

14.4 Additional surveillance visits may be required if substantiated complaints against the *Assessed Enterprise* have been received by the *Certification* or *Registration Body*.

15. TECHNICAL *ASSESSMENT*

15.1 An *Enterprise* shall be required to make available for inspection *Electrical installation work*, completed or in progress, representative of the category(s) of work to which an application for, or existing, certification or registration relates, and sufficient for the *Assessment* process. The required technical standard of the work shall not be less than that detailed in the relevant technical reference documents listed in Appendix 2.

8

A

EAS 15-350

15.2   For the purposes of an application for certification or registration, such electrical work shall have been carried out by the *Enterprise* within the 12 months prior to the application, in each category for which certification or registration is sought.

15.3   The *Certification* or *Registration Body* shall require the *Enterprise* to demonstrate safe electrical isolation procedure.

15.4   An *Enterprise* shall be required to permit representatives of the *Certification* or *Registration Body* to have access to the *Contracting Office* in order to assess equipment, documentation and related business systems.

15.5   The extent of *Assessment* will be prescribed by the *Certification* or *Registration Body* having regard to the range, scale and geographical spread of *Electrical installation work* for which *Assessment* is sought.

15.6   The *Enterprise* shall be required to have the following items available for *Assessment* by the *Certification* or *Registration Body*'s representative together with any other items as prescribed and published from time to time by the *Certification* or *Registration Body*:

15.6.1   technical reference documents (Appendix 2);

15.6.2   test instruments, appropriate to the range of *Electrical installation work* undertaken, including a record of *Assessment* of accuracy (Appendix 3);

15.6.3   a record of all *Electrical installation work* in progress and completed over the previous 12 months or since the previous *Assessment*, whichever is the longer period;

15.6.4   specifications, drawings, certificates and reports relating to work in progress and completed over the previous 12 months or since the previous *Assessment*, whichever is the longer period;

15.6.5   any other items the *Certification* or *Registration Body* requires which are relevant to the *Assessment* process;

15.6.6   evidence that the required public liability insurance cover is held;

15.6.7   a record of all complaints received over the previous three years about the technical standard, safety and *Functionality* of *Electrical installation work*, and details of actions taken to resolve the complaints;

15.6.8   evidence that a written health and safety policy statement is in place and that risk *Assessment*s are carried out as appropriate.

15.7   At the time of the *Assessment*, the *Certification* or *Registration Body* shall also review the complaints record to ascertain whether there have been failures of installed systems to operate as required by the relevant standard(s) listed in Appendix 2.

15.8   The *Enterprise* shall be required to be fully prepared for the *Assessment* by the *Certification* or *Registration Body*'s representative, including making the items required to be seen readily available. The *Qualified Supervisor*(s) for the category(s) of work for which the *Enterprise* is to be assessed shall be required to be present throughout the *Assessment* process and the *Principal Duty Holder* shall be required to be available to discuss the intent and result of the *Assessment*.

9

15.9   Registered and *Proposed Qualified Supervisor*s shall be subject to *Assessment* in accordance with the requirements set out in Section 11.

15.10 The *Enterprise* shall be required to provide facilities and shall arrange access to sites to inspect *Electrical installation work* selected for *Assessment* by the *Certification* or *Registration Body*.

16.   RECORDS

16.1   In addition to those records detailed previously, the *Assessed Enterprise* shall be required to hold for at least three years a list of all *Electrical installation work* carried out together with the specifications, drawings, certificates and other relevant documents relating to that work.

17.   *ASSESSMENT* DECISION

17.1   On completion of the *Assessment* by the representative of the *Certification* or *Registration Body*, the *Enterprise* shall receive an *Assessment* report, recording all observed *non-conformities*.

17.2   The *Enterprise* shall subsequently be advised by the *Certification* or *Registration Body* of its decision as to whether certification or registration is to be granted.

18.   ASSESSMENT CERTIFICATE

18.1   A *Contracting Office* of an *Enterprise* having been assessed as complying with the requirements of this Specification and being in possession of a current *Assessment Certificate* shall be permitted to advertise its services as an Assessed *Enterprise*.

18.2   The Certificate shall be issued by the *Certification* or *Registration Body* for each *Contracting Office* assessed stating that the *Contracting Office* has achieved compliance with this Specification in respect of the range and scope of work assessed.

18.3   The Certificate shall remain the property of the *Certification* or *Registration Body* and shall be required to be returned, upon request, on cessation of certification or registration for whatever reason.

18.4   The *Assessed Enterprise* shall, at all reasonable times on request, be required to produce its *Assessment Certificate* to a representative of the *Certification* or *Registration Body*.

18.5   In being granted certification or registration, the *Assessed Enterprise* shall be required to undertake to continue to comply with the requirements of this Specification for the period for which the *Assessment Certificate* remains valid.

18.6   If, during a surveillance visit, the *Assessed Enterprise* is unable to offer sufficient examples of work for evaluation purposes in a category included in its scope of certification, the *Certification* or *Registration Body* may suspend, cancel or reduce the scope of certification or registration accordingly.

18.7   The *Assessed Enterprise* shall be required not to claim competence under this Specification for categories of work other than those for which it holds a current *Assessment Certificate*.

19. CHANGE OF DETAILS

19.1   An *Assessed Enterprise* shall be required to give notice to the *Certification* or *Registration Body* of a change of legal constitution, trading or other title, address, *Principal Duty Holder*, *Qualified Supervisor* or other significant particulars and declarations upon which the current *Assessment Certificate* was granted. Such notice shall be required to be given to the *Certification* or *Registration Body* within thirty days of any such change becoming effective. From the date any such change affecting the *Qualified Supervisor* occurs, the *Assessed Enterprise* shall be required to have in place within a period of 120 days a replacement who has been assessed by the *Certification* or *Registration Body* as competent for the category(s) of work undertaken.

19.2   Where, in the opinion of the *Certification* or *Registration Body*, the changes are such that the conditions under which an *Assessment Certificate* was granted are significantly affected, a new application for certification or a surveillance visit may be required.

19.3   An *Assessed Enterprise* shall be eligible to remain certified or registered for the period covered by the *Assessment Certificate* provided it continues to be engaged in *Electrical installation work* in the category(s) for which it holds an *Assessment Certificate* and continues to comply with this Specification.

20. CANCELLATION OF *ASSESSMENT CERTIFICATE*

20.1   An *Assessment Certificate* shall be subject to cancellation or amendment by the *Certification* or *Registration Body* if an Assessed *Enterprise*:

20.1.1   makes wilful misrepresentation in its application for certification or registration;

20.1.2   fails to complete, to the satisfaction of the *Certification* or *Registration Body*, the remedial action it requires as a result of a customer complaint being dealt with under Section 22;

20.1.3   culpably or negligently creates or causes to be created danger or serious risk of injury through the use, in *Electrical installation work*, of faulty materials, materials *not conforming* to recognised standards, or through faulty workmanship;

20.1.4   becomes bankrupt or insolvent or having a Receiving Order made against it or compounds with its creditors or being a corporation commences to be wound up (not being a members' voluntary winding up for the purposes of reconstruction) or carries on business under a Receiver for the benefit of its creditors or any of them or if in the opinion of the *Certification* or *Registration Body* the nature of its work has changed or it shall cease to trade or if there be any change in the ownership of its *Enterprise* which affects the conditions under which it was certified;

20.1.5   claims to have been certified for *Electrical installation work* not included at the time in the scope of its *Assessment Certificate*.

20.2   An *Assessment Certificate* may be subject to cancellation or amendment by the *Certification* or *Registration Body* if an Assessed *Enterprise*:

20.2.1   commits a breach of any of the obligations imposed by this Specification;

20.2.2   undertakes *Electrical installation work* below the technical standard required or is unable to continue to comply with this Specification;

11

20.2.3   performs any act which, in the opinion of the *Certification* or *Registration Body*, is contrary or prejudicial to the objects or reputation of the *Certification* or *Registration Body*;

20.2.4   makes use of a *Certification* or *Registration Body*'s approved mark or logo in a manner which, in the opinion of the *Certification* or *Registration Body*, is likely to bring the *Certification* or *Registration Body* into disrepute;

20.2.5   uses its certification or registration in a manner as to bring the *Certification* or *Registration Body* into disrepute or makes any statement regarding its certification or registration which the *Certification* or *Registration Body* may consider to be misleading.

20.3   The *Certification* or *Registration Body* shall notify an *Assessed Enterprise* in writing of an intention to cancel its *Assessment Certificate*, detailing fully such reasons for its action. If the *Assessed Enterprise* wishes to object, it shall be required to notify the *Certification* or *Registration Body* in writing, within twenty-one days, of its objections, for consideration by the *Certification* or *Registration Body*.

20.4   On cancellation of an Assessed *Enterprise*'s *Assessment Certificate*, the requirements of Clause 23.2 shall apply.

21.   APPEALS, COMPLAINTS AND DISPUTES

21.1   The *Certification* or *Registration Body* shall have procedures in place to deal with any appeals, complaints and disputes from an Assessed *Enterprise*. Such procedures shall be made available to *Enterprise*s.

21.2   An *Enterprise* shall be permitted to appeal against any decision of the *Certification* or *Registration Body* in respect of its certification or registration.

21.3   The *Enterprise* shall be required to give notice in writing setting out clearly the grounds for such an appeal. Such an appeal shall be required to be served on the *Certification* or *Registration Body* within twenty-one days of the date of notification of the decision.

21.4   The *Enterprise* and the *Certification* or *Registration Body* shall be required to bear their own costs associated with any appeal, regardless of the outcome.

22.   COMPLAINTS ABOUT AN *ASSESSED ENTERPRISE*'S *ELECTRICAL INSTALLATION WORK*

22.0   The *Certification* or *Registration Body* shall have procedures in place to deal with complaints in a timely manner.

22.1   Where a complaint against an *Assessed Enterprise* cannot be resolved by the *Enterprise* and the complaint is subsequently received by a *Certification* or *Registration Body* indicating that the work undertaken by the *Assessed Enterprise*, or sub-let by the *Assessed Enterprise*, does not meet the requirements of the relevant standards listed in Appendix 2, the *Assessed Enterprise* shall be required to provide facilities for inspections to be carried out by the *Certification* or *Registration Body*, including test equipment and access to the work to be inspected.

22.3   Where, as a result of such inspections, it is shown to the satisfaction of the *Certification* or *Registration Body* that the standard of the work is below that required; the *Assessed Enterprise* shall be required, at its own expense, to take remedial action within a specified time as notified to it by the *Certification* or *Registration Body*.

12

A

22.4 The *Certification* or *Registration Body* shall investigate complaints about an *Assessed Enterprise*'s *Electrical installation work* where:

22.4.1 the work was undertaken within the scope of the *Assessment Certificate*; and

22.4.2 the work was undertaken within the last two years or was undertaken since the date of issue of the *Assessment Certificate*, whichever is the shorter period.

23. *ASSESSMENT* MARKS AND LOGOS

23.1 If the *Certification* or *Registration Body* confers the right to use its mark or logo to indicate certification or registration in accordance with this Specification, the *Assessed Enterprise* shall be required to use the specified mark or logo only as authorised in writing by the *Certification* or *Registration Body*.

23.2 Upon cancellation of its *Assessment Certificate*(s), however determined, the *Enterprise* shall be required to immediately discontinue use of all advertising matter, stationery etc. containing reference to certification or registration and return *Assessment Certificate*s as required by the *Certification* or *Registration Body*.

13

C) APPENDICES

Appendix 1 Work categories

A   Electrical installations up to 1kV

A1 Electrical installations
A1.1   *Dwelling*s
A1.2   All other
A1.3   Temporary installations as defined in Appendix 9 |

Appendix 2 Technical reference documents

For all categories of work under section A of Appendix 1, the *Enterprise* shall, as a minimum, be required to hold the current version (including all amendments) of:

- BS 7671: Requirements for Electrical Installations
- Memorandum of Guidance on the Electricity at Work Regulations (HSR25)
- All building regulations approved documents relevant to the work undertaken by the *Enterprise*.

For categories of work under Section A1.3 of Appendix 1, additional technical reference documents are given in Appendix 9.

This list is not exhaustive. Other requirements may be set by the *Certification* or *Registration Body* as may be required for the work undertaken.

© The Institution of Engineering and Technology

Appendix 3 Test instruments – Calibration requirements

The *Enterprise* shall be required to have a suitable system in place to ensure that the accuracy and consistency of all test instruments used for certification and reporting purposes is being maintained.

There are a number of alternatives for such control systems, including:

- Maintaining records of the formal calibration/re-calibration of test instruments as recommended by the instrument manufacturers, supported by calibration certificates with measurements traceable to national standards, issued by organizations recognized by *Certification* or *Registration Bodies* for the purposes of checking the accuracy of test instruments. Certificates issued by UKAS accredited laboratories are preferable

- Maintaining records over time of comparative cross-checks with other test instruments used by the *Enterprise*

- Maintaining records over time of measurements of the characteristics of designated reference circuits or devices. For example, the consistency of continuity, insulation resistance and earth electrode test instruments could be checked against a proprietary resistance box or a set of suitable resistors. Earth fault loop impedance test instruments could be checked by carrying out tests on a designated socket-outlet (on a non-RCD protected circuit) in the *Enterprise*'s office. RCD test instruments could be checked by carrying out tests on an RCD unit plugged into the designated socket-outlet.

For all low voltage *Electrical installation work*, the *Enterprise* shall be required to hold the following test instruments as a minimum:

- Insulation resistance test instrument
- Continuity test instrument
- Voltage indicating instrument*
- Earth fault loop impedance test instrument
- Residual current device test instrument
- Suitable split test leads for both the phase/earth loop impedance test instrument and the residual current device test instrument.

Two or more of the functions of the above test instruments may be combined in a single instrument.

The *Certification* or *Registration Body* may vary the requirements for an *Enterprise* that only undertakes extra-low voltage *Electrical installation work*.

In addition, the *Enterprise* shall be required to hold additional test instruments particular to the scope of work being assessed.

  * Voltage indicating equipment does not require calibration.

16

A

Appendix 4 Requirements for the registration of *Qualified Supervisors*

The Criteria in this appendix apply to applications made after 6th April 2013, they are not retrospective.

TABLE 4A MINIMUM TECHNICAL COMPETENCE CRITERIA REQUIRED FOR A PROPOSED QUALIFIED
SUPERVISOR FOR ELECTRICAL INSTALLATIONS IN DWELLINGS ONLY

Entry route	Knowledge and Understanding Requirements	Experience
1	Level 3 Certificate in Installing, Testing and Ensuring Compliance of Electrical installations in *Dwellings*	Must provide evidence of work carried out to be able to demonstrate their competence for the scope for which they have applied.
	Equivalents	
2	NVQ 3 Electrotechnical Services (Installation, Buildings and Structures) Plus Level 3 Award in the Initial Verification and Certification of Electrical Installations Or Level 3 Award in Approving *Electrical Installation Work* in *Dwellings* in Compliance with Building Regulations And Current Level 3 Award in the Requirements for Electrical Installations	Must provide evidence of work carried out to be able to demonstrate their competence for the scope for which they have applied.
3	Relevant electrical installation *Qualification* preceding NVQ Plus Level 3 Award in the Initial Verification and Certification of Electrical Installations Or Level 3 Award in Approving *Electrical Installation Work* in *Dwellings* in Compliance with Building Regulations And Current Level 3 Award in the Requirements for Electrical Installations	Must provide evidence of work carried out to be able to demonstrate their competence for the scope for which they have applied AND evidence of ongoing Continuous Professional Development.
4	Auditable evidence, for example *Certificates of competence,* that reflect the learning outcomes identified in the Level 3 Certificate in Installing, Testing and Ensuring Compliance of Electrical installations in Dwellings based on the *National Occupational Standards.*	Must provide evidence of work carried out to be able to demonstrate 2 years' experience for the scope for which they have applied.
5	Existing or previously recognised *Qualified Supervisors* registered within last two years with the Current Level 3 Award in the Requirements for Electrical Installations	Letter or similar from the previous *Scheme* provider confirming previous status and must provide evidence of work carried out to demonstrate practical competence for the scope for which they have applied.

17

TABLE 4B MINIMUM TECHNICAL COMPETENCE CRITERIA REQUIRED FOR A PROPOSED QUALIFIED SUPERVISOR FOR ALL ELECTRICAL INSTALLATION WORK*

Entry route	Knowledge and Understanding Requirements	Experience
1	**Level 3 NVQ Diploma in Installing Electrotechnical systems and equipment (building structures and the environment)** Plus Level 3 Award in the Periodic Inspection, Testing and Certification of Electrical Installations	Must provide evidence of work carried out to be able to demonstrate 2 years of competence for the scope for which they have applied.
	Equivalents	
2	**NVQ 3 Electrotechnical Services (Installation, Buildings and Structures)** Plus Level 3 Award in the Initial Verification and Certification of Electrical Installations And Level 3 Award in the Periodic Inspection, Testing and Certification of Electrical Installations And Current Level 3 Award in the Requirements for Electrical Installations	Must provide evidence of work carried out to be able to demonstrate 2 years competence for the scope for which they have applied
3	**Relevant electrical installation *Qualification* preceding NVQ** PLUS Level 3 Award in the Initial Verification and Certification of Electrical Installations And Level 3 Award in the Periodic Inspection, Testing and Certification of Electrical Installations And Current Level 3 Award in the Requirements for Electrical Installations	Must provide evidence of work carried out to be able to demonstrate 2 years competence for the scope for which they have applied AND evidence of ongoing Continuous Professional Development.
4	**Level 3 Certificate in Installing, Testing and Ensuring Compliance of Electrical installations in *Dwellings*** Plus **The additional knowledge and understanding requirements of route 1 (as above).**	Must provide evidence of work carried out to be able to demonstrate 2 years' experience for the scope for which they have applied.
5	**Auditable evidence, for example *Certificates of competence*, that reflects the units of competence identified in the Level 3 NVQ Diploma in Installing Electrotechnical systems and equipment (building structures and the environment) based on the *National Occupational Standard*.**	Must provide evidence of work carried out to be able to demonstrate 2 years competence for the scope for which they have applied.
6	**Existing or previously recognised *Qualified Supervisor*s registered within last two years** with the Current Level 3 Award in the Requirements for Electrical Installations	**Letter or similar from the previous *Scheme* provider confirming previous status and must provide evidence of work carried out to demonstrate practical competence for the scope for which they have applied.**

* For those involved in temporary *electrical installation work* for entertainment or events, as defined in Appendix 9,
route 5 can be augmented by "C&G 181 Entertainment and Theatre electrician" or equivalent.

18

A

Appendix 5 *Assessment certificate*

An *Assessment Certificate* shall, as a minimum, contain the following information:

Trading title and address of the *Contracting Office* of the *Enterprise*

Registration number of the *Enterprise*

Date of issue

Date of expiry (if any)

Category(s) of work for which the *Enterprise* has been assessed*

Name and address of *Certification* or *Registration Body*

A statement to the effect that the validity of the certificate can be checked by consulting the awarding *Certification* or *Registration Body*, together with details of how the validity can be checked.

* The Certificate shall include, or be supplemented by, information detailing the exact extent or limitations of the *Electrical installation work* within the scope of the Certificate.

19

Appendix 6 Particular requirements for building regulations *competent person scheme operator*s covering *Electrical installation work* in *dwelling*s

1.   **The requirements set out in this Appendix vary the basic *Criteria* in the body of the Specification.**

2.   This Appendix sets out the *Criteria* to be used by a body offering a self-certification *Scheme* covering *Electrical installation work* in *dwelling*s listed in Schedule 3 of Building Regulations 2010, as amended, in assessing technical competence and the capability of an *Enterprise* to undertake *Electrical installation work* in *dwelling*s prior to registration.

3.   This Appendix defines the requirements in relation to two levels of activity for *Enterprise*s:

3.1   **Full scope:** The requirements for an *Enterprise* the scope of whose work includes the design, installation, inspection and testing of *Electrical installation work* that is associated with *dwelling*s and is intended to operate at low or extra-low voltage

3.2   **Defined scope:** The requirements for an *Enterprise* the scope of whose work is limited to the design, installation, inspection and testing of defined *Electrical installation work* that is associated with *dwelling*s, is intended to operate at low or extra-low voltage, and is undertaken in connection with, or ancillary to some other work.

DEFINITIONS

Competent Person Scheme Operator - a body offering a self-certification *Scheme* covering *Electrical installation work* in *dwelling*s and meeting the requirements of this document and listed in Schedule 3 of the Building Regulations 2010, as amended.

Dwelling – a house or a *Flat*.

Flat – Separate and self-contained premises constructed or adapted for use for residential purposes and forming part of a building from some other part of which it is divided horizontally.

Responsible Person – In respect of a *Defined Scope Enterprise*, a *Competent Person* with specific responsibility on a day-to-day basis for the safety, technical standard and quality of *Electrical installation work* in *dwelling*s.

Registered Responsible Person – a *Responsible Person* who has been assessed and accepted by the *Certification* or *Registration Body*.

GENERAL REQUIREMENTS FOR *FULL SCOPE* AND *DEFINED SCOPE ENTERPRISE*S

4.   ***Electrical installation work***

4.1   An *Assessed Enterprise* shall be required not to sub-let *Electrical installation work* in *dwelling*s unless the work is undertaken by an *Enterprise* that is registered with a *Competent Person Scheme Operator*. *Electrical installation work* shall be required to be self-certified only by the *Assessed Enterprise* that carried out that work.

4.2   Competence of the *Enterprise* will not be assessed on *Electrical installation work* sub-let to others. However the *Assessed Enterprise* will be required to retain responsibility for this work.

5. Certification and Reporting

5.1 The *Assessed Enterprise* shall be required to issue the appropriate Building Regulations compliance certificates in accordance with the relevant standards and regulations (see 8.1 below) for all the *Electrical installation work* in *dwelling*s that it carries out.

5.2 The *Assessed Enterprise* shall be required to hold for at least three years copies of all the certificates that it has issued for *Electrical installation work* in *dwelling*s.

6. Complaints

6.1 The *Assessed Enterprise* shall be required to maintain a record of all complaints received over at least the previous three years, concerning the compliance with Building Regulations of the *Electrical installation work* it has carried out in *dwelling*s, together with the details of the actions taken to resolve these complaints.

7. Health and Safety

7.1 The *Enterprise* shall be required to have a written health and safety policy statement, and shall be required to carry out risk *Assessment*s as appropriate.

8. Technical Reference Documents

8.1 *Enterprise*s shall, as a minimum, be required to hold the current version (including all amendments) of:

Technical Reference Document	*Full scope*	*Defined scope*
BS 7671 Requirements for Electrical Installations	Yes	Not essential
IET On-Site Guide to BS 7671 or equivalent.	Not essential	Yes
HS(R)25 Memorandum of Guidance on the Electricity at Work Regulations	Yes	Yes
Building Regulations, Approved Document P.	Yes	Yes

21

9. Testing

9.1   Test instruments and suitable leads are required for *Full scope Enterprises* to carry out the tests in the table below to comply with Section 9 of the Specification.

Test
Voltage measurement
Insulation resistance measurement
Continuity
Polarity
Measurement of earth fault loop impedance of circuits
Measurement of external earth fault loop impedance
Operation of residual current devices

9.2   For *Defined scope Enterprises*, test instruments and suitable leads are required to carry out the relevant tests in relation to the defined *Electrical installation work* of the *Enterprise* to comply with Section 9 of this Specification.

FULL SCOPE ENTERPRISES: PARTICULAR REQUIREMENTS RELATING TO THE COMPETENCE OF AN *ENTERPRISE* TO DESIGN, INSTALL, INSPECT AND TEST FIXED ELECTRICAL INSTALLATIONS IN *DWELLINGS*.

10   Responsibilities of a *Qualified Supervisor*:

10.1   Ensuring compliance of the *Electrical installation work* with the relevant requirements of the Building Regulations.

10.2   Verifying and authenticating certification showing compliance with Building Regulations.

11   Requirements for a *Qualified Supervisor*

11.1   A *Qualified Supervisor* shall be required to demonstrate the technical competence to be registered as a *Qualified Supervisor* so as to ensure the compliance of the *Electrical installation work* in *dwelling*s carried out by the *Enterprise* with the Building Regulations (England & Wales).

DEFINED SCOPE ENTERPRISES:
PARTICULAR REQUIREMENTS RELATING TO THE COMPETENCE OF AN *ENTERPRISE* TO UNDERTAKE DEFINED *ELECTRICAL INSTALLATION WORK* IN *DWELLINGS* IN CONNECTION WITH, OR ANCILLARY TO, SOME OTHER NON-ELECTRICAL WORK.

12.   Defined *Electrical installation work*

12.1   The *Enterprise* shall be directly engaged in carrying out defined *Electrical installation work* in *dwelling*s.

12.2   Defined *Electrical installation work* is electrical work that is provided solely within the scope of *Electrical installation work* defined for the *Enterprise*.

Electrician's Guide to the Building Regulations | **217**
© The Institution of Engineering and Technology

13.   Personnel

13.1   *Responsible Person*s shall be subject to the same requirements as those for *Qualified Supervisor*s set out in the body of this Specification.

13.2   The *Enterprise* shall be required to appoint as *Responsible Person*s as many individuals as necessary to ensure all *Electrical installation work* in *dwelling*s is carried out by or under the supervision of a *Responsible Person*. The *Principal Duty Holder* may also be a *Responsible Person*.

13.3   *Responsible Person*s shall be required to hold the responsibilities detailed in Section 14 of this Appendix, and shall be required to meet the requirements set out in Section 15 of this Appendix.

13.4   Where a *Registered Responsible Person* ceases to be employed in that capacity, the *Principal Duty Holder* shall be required to immediately notify the *Certification* or *Registration Body*.

13.5   A *Principal Duty Holder* shall ensure that all *Electrical installation work* in *dwelling*s is allocated to the *Enterprise*'s *Responsible Person*(s).

13.6   A *Principal Duty Holder* shall ensure that *Electrical installation work* in *dwelling*s that goes beyond the *Enterprise*'s *Defined scope* of *Electrical installation work*, as described in the scope, is referred back to the customer for execution by an *Enterprise* that has the necessary scope.

14   Responsibilities of a *Responsible Person*:

14.1   The safety, quality and technical standard on a day-to-day basis of the *Electrical installation work* in *dwelling*s allocated to him/her (or where appropriate other persons under his supervision), by the *Principal Duty Holder*

14.2   Ensuring compliance of the *Electrical installation work* with the relevant requirements of the Building Regulations.

14.3   Ensuring that the results of the inspection and testing are accurately recorded on the appropriate forms of certification.

14.4   Verifying and authenticating certification showing compliance with Building Regulations.

15   Requirements for a *Responsible Person*

15.1   A *Responsible Person* shall be required to:

15.1.1 have adequate knowledge, experience and understanding of the design, installation, inspection and testing procedures for *Electrical installation work* in *dwelling*s in accordance with BS 7671 for each of the relevant competence requirements set out in Appendix 4 above and;

15.1.2 demonstrate the technical competence to be registered as a *Responsible Person* so as to ensure the compliance of the *Electrical installation work* in *dwelling*s carried out by the *Enterprise* with the Building Regulations (England & Wales).

16   *Registered Responsible Person*s shall be required to hold an appropriate BS 7671 *Qualification* awarded by an *Accredited body* within two years of a change to the BS 7671 regulations coming into effect or be able to demonstrate an equivalent level of knowledge.

23

A

EAS 15-350

Appendix 7 Particular requirements for Scottish Building Standards

DEFINITIONS

Approved Body – A contractor or other *Enterprise* registered in a Certification *Scheme* to employ and support Approved Certifiers of Construction.

Approved Certifier of Construction – An individual registered in a Certification *Scheme* to determine compliance of electrical installations with the Scottish Building Standards Technical Handbooks and provide Certificates of Construction.

Scheme Provider – An organisation that operates one or more Certification *Scheme*s and registers Approved Certifiers of Design and/or Construction and Approved Bodies

1. The requirements set out in this appendix vary the base *Criteria* in the body of the specification and summarise the *Criteria* for registration in *Scheme*s for Approved Certifiers of Construction as provided for in the Building (Scotland) Act 2003. These *Scheme*s allow qualified and experienced trades-people to be responsible for ensuring compliance with the Building (Scotland) Regulations as amended without detailed scrutiny by local authority verifiers. These Approved Certifiers must be employed by reputable organisations and be able to demonstrate that they meet the detailed *Criteria* to be registered in a relevant *Scheme*.

2. Persons who perform the function of certification are known as Approved Certifiers of Construction. To register and operate as an Approved Certifier an individual must be employed by an *Approved Body* that has gained entry to an appropriate certification *Scheme*.

3. An *Approved Body* is an *Enterprise* that employs and supports one or more Approved Certifiers, and meets the conditions for registration in a certification *Scheme* operated by a *Scheme Provider*.

4. The scope of the *Scheme* approved by Scottish Ministers is the certification, installation and commissioning of electrical installations to BS 7671 as complying with Technical Standards

5. Approved Certifiers of Construction

    5.1 The function of an *Approved Certifier of Construction* is to certify that an electrical installation to BS 7671 complies with the current Building (Scotland) Regulations as amended.

    5.2 The responsibility of an *Approved Certifier of Construction* is to be satisfied that their knowledge and experience enables them to discharge the responsibility of certifying particular work, bearing in mind the nature of the project. If appropriate, the *Approved Certifier of Construction* should call on other suitably qualified and experienced persons for advice. Certification is however the sole responsibility of the *Approved Certifier of Construction* who signs the Certificate of Construction.

24

A

6. Requirements for an *Approved Certifier of Construction*

6.1 Individual registration with the *Scheme* is open to any person who is eligible to qualify for a Scottish Joint Industry Board (SJIB) Approved Electrician Grade Card and is a principle or full-time employee of an *Approved Body*. In addition the individual should:

- have evidence of having passed an assessed course in all relevant requirements in the Scottish Building Standards technical handbooks and current Building (Scotland) Regulations as amended within the previous three years; and
- have evidence of a level 3 Certificate in the Requirements for Electrical Installations to BS 7671:2008 awarded by a body regulated by the Office of Qualifications and Examinations Regulation (OFQUAL) or the Scottish Qualifications Authority (SQA) or having passed an assessed training course acceptable for grading purposes by the SJIB within the previous five years; and
- can demonstrate current knowledge of the Scottish Building Standards System and a sufficient understanding of the role of *Approved Certifiers of Construction* (Electrical installations to BS 7671).[1]

7. Approved Bodies

7.1 The functions of an *Approved Body* are to support the Approved Certifiers of Construction employed or contracted by them and to maintain the conditions under which the *Approved Body* was approved.

7.2 Registration with the *Scheme* is open to any *Enterprise* that is directly engaged in *Electrical installation work*, has carried out such work for not less than six months, employs at least one *Approved Certifier of Construction* and undertakes to comply with the rules, which include but are not limited to:

- Appointing a Certification Co-ordinator to counter-sign Certificates completed by an *Approved Certifier of Construction*;
- Offering certification services under the *Scheme* only if it employs at least one *Approved Certifier of Construction* and a Certification Co-ordinator
- Is directly engaged in certifying *Electrical installation work* for compliance with Scottish Building Regulations;
- Operates an appropriate system of checking compliance with Scottish Building Regulations in accordance with *Scheme Providers*' requirements to ensure the quality of certification activities;
- Completes Certificates of Construction in the form prescribed by *Scheme Providers*, using certificate numbers designated by *Scheme Providers*. All Certificates of Construction shall be signed by an *Approved Certifier of Construction* and counter-signed by the Certification Co-ordinator;
- Holds records for at least five years and makes them available to the *Scheme Provider* for audit when required. Records shall include check-lists, Certificates of Construction and training and development of Approved Certifiers of Construction.

---

[1] It is important to understand that the criteria listed for Approved Certifiers and Bodies in this Appendix are a summary. A full understanding of the detailed criteria and the Scheme in its entirety is essential for organisations and individuals wishing to work in Scotland. A comprehensive guide to the Scheme for Certification of Construction (Electrical Installations to BS 7671), in accordance with the requirements of the Building (Scotland) Act 2003 and the current Building Procedure (Scotland) Regulations is available at www.sbsc.uk.net

25

8   Sole traders

8.1 A sole trader may hold membership of the *Scheme* as an *Approved Certifier of Construction*, an *Approved Body* and also take the role of Certification Co-ordinator.

Appendix 8 Interpretations of BS EN 45011*1 in relation to the specification

1   INTRODUCTION

1.1   The general requirements for accreditation are laid down in the European Standard, BS EN 45011*1 - *General requirements for bodies operating product certification systems*. The word 'product' includes a process or service. The requirements of BS EN 45011*1 apply to all types of product, process and service certification, and therefore need to be interpreted in respect of the *Assessment* of the competency of *Enterprises* based on this Specification.

1.2   This Appendix in conjunction with European co-operation for Accreditation document EA-6/01 is intended to amplify or clarify those requirements of BS EN 45011*1 that need interpretation when applied to this Specification. This Appendix does not cover all the requirements of BS EN 45011*1, and *Certification Bodies* are reminded of the need to comply with all of the requirements of that standard. Cross-reference is made to relevant sections of BS EN 45011*1.

1.3   The Accreditation Body in the UK, the United Kingdom Accreditation Service (UKAS), will apply the interpretations given in this Appendix when assessing and re-assessing *Certification Bodies*. In cases of difficulties of interpretation arising from this Appendix, the EAS Management Committee will provide clarification.

5   SCOPE (BS EN 45011*1, Section 1)

5.1 (1.1)   The scope of BS EN 45011*1 includes the certification of a service, such as the carrying out of *Electrical installation work* within the scope of this Specification.

5.2 (1.2)   This Appendix provides interpretations concerning the application of BS EN 45011*1 to *Schemes* based on this Specification.

6   REFERENCES (BS EN 45011*1, Section 2)

Add the following reference:

6.1   Electrotechnical *Assessment* Specification for use by *Certification* and *Registration Bodies*

7   CERTIFICATION BODY (BS EN 45011*1, Section 4)

7.1   General provisions (BS EN 45011*1, Subsection 4.1)

7.1.1 (4.1.3) The *Criteria* to be used by the *Certification Body* in carrying out an evaluation shall be those set out in this Specification. If explanation is required as to the application of those *Criteria* to a particular *Assessment Scheme*, it shall be formulated by the body responsible for developing the Specification.

7.2   Organisation (BS EN 45011*1, Subsection 4.2)

7.2.1 (4.2g) Any *Certification Body* wishing to base an *Assessment Scheme* on the Specification must be accredited to BS EN 45011*1 taking into account the interpretations given in this Appendix, and must carry out *Assessments* against the requirements of this Specification.

7.2.2 (4.2n) Decisions on certification shall be taken under the supervision of a certification committee on which there is a balance of interests.

27

7.3 Subcontracting (BS EN 45011*1, Subsection 4.4)

7.3.1 (4.4b) A body undertaking *Assessment*s under an *Assessment Scheme* based on the Specification as a subcontractor to an accredited *Certification Body* need not itself be accredited to BS EN 45011*1 under this Specification, but shall comply in all other respects with the relevant requirements of this Specification.

7.4 Quality system (BS EN 45011*1, Subsection 4.5)

7.4.1 (4.5.2) The quality system to be operated by the *Certification Body* shall be in accordance with the relevant elements of BS EN 45011*1 and of this Specification.

7.5 Conditions and procedures for granting, maintaining, extending and withdrawing certification (BS EN 45011*1, Subsection 4.6)

7.5.1 (4.6.1) The conditions for granting, maintaining, extending and withdrawing certification are set out this Specification.

7.6 Documentation (BS EN 45011*1, Subsection 4.8)

7.6.1 (4.8.1) In providing information about its certification system, the *Certification Body* shall include reference to this Specification.

7.6.2 (4.8.2) The *Certification Body* shall hold a copy this Specification.

7.6.3 (4.8.1e) In describing the rights and duties of *Assessed Enterprise*s, *Certification Bodies* shall refer to this Specification.

7.6.4 (4.8.1f) In providing information about the handling of complaints, appeals and disputes, *Certification Bodies* shall refer to this Specification.

7.6.5 (4.8.1g) The directory of *Assessed Enterprise*s published by the *Certification Body* shall provide at least the following information for each *Contracting Office* of an *Assessed Enterprise*:

7.6.5.1 Registered trading title of the *Assessed Enterprise*

7.6.5.2 Full address of the *Contracting Office*

7.6.5.3 Telephone number of the *Contracting Office*

7.6.5.4 The categories and types of work for which the *Contracting Office* holds an *Assessment Certificate*

7.6.6 The information shall be updated and published by the *Certification Body* at intervals of no more than twelve months.

7.7 Records (BS EN 45011*1, Subsection 4.9)

7.7.1 (4.9.1) The *Certification Body* shall maintain a record system to suit the *Assessment* procedures set out in this Specification.

8 *CERTIFICATION BODY* **PERSONNEL** (BS EN 45011*1, SECTION 5)

8.1 *Qualification Criteria* (BS EN 45011*1, Subsection 5.2)

28

8.1.1 (5.1.1) Personnel involved in the *Assessment* process shall have *Qualifications* and experience at least to the minimum requirements set out in this Specification, supplemented by appropriate training as an Assessor.

9   CHANGES IN CERTIFICATION REQUIREMENT

(6)   Any changes in certification requirements should not conflict with the requirements of this Specification.

10   APPEALS, COMPLAINTS AND DISPUTES

10.1 (7.1)   Appeals, complaints and disputes must take into account the requirements of this Specification.

10.2 (7.1)   A *Certification Body* shall have in place a system for recording and, where reasonably practicable, resolving complaints against *Assessed Enterprise*s notified to the *Certification Body*.

10.3 (7.2)   Due account must be taken of the requirements given in this Specification.

11   APPLICATION FOR CERTIFICATION

11.1 (8.1)   Information on the procedure:

11.1.1 (8.1.1)   The *Certification Body* should include in its *Assessment* procedures the minimum requirements set out in this Specification.

11.1.2 (8.1.2b)   Certification will not be awarded for categories of work where there is no evidence that the *Enterprise* has been undertaking that work.

11.1.3 (8.1.2c)   Scope means as per this Specification.

11.1.4 (8.1.2d)   This will be as required by this Specification.

11.1.5 (8.1.2e)   There is no provision for suspension in this Specification. Any *Certification Body* may have procedures that allow suspension and shall allow appeals in line with Section 21 of this Specification.

11.1.6 (8.1.2h)   The supplier shall also ensure that the requirements of this Specification are met.

11.2 (8.2)   The application:

11.2.1 (8.2.1)   The application for certification shall be signed by the *Principal Duty Holder*. The application shall also be made in accordance with Section 13 of this Specification.

11.2.2 (8.2.2b)   The category(s) of work applied for shall also be provided.

29

12   PREPARATION FOR EVALUATION

12.1 (9.3)   Personnel undertaking the *Assessment* of *Enterprises* shall meet the *Qualification* requirements of this Specification.

12.2 (9.4)   *Certification Body* personnel shall be provided with a copy of this Specification and the technical standards relevant to the category(s) of work for which the applicant is applying for *Assessment*.

13   EVALUATION

13.10.1   The *Certification Body* shall need to evaluate a representative sample of work undertaken for each category of work applied for.

14   EVALUATION REPORT

14.11.1   Each report submitted to the *Certification Body* shall be evaluated to ensure its accuracy. The person undertaking the evaluation should not have been involved in the compilation of the original report.

15   DECISION ON CERTIFICATION

15.1 (12.3)   The Certificate issued to the applicant shall comply with the requirements as shown in Appendix 5.

15.2 (12.4)   The *Certification Body* shall provide an amendment to the scope of certification in accordance with this Specification.

16   SURVEILLANCE

16.1 (13.4)   The *Certification Body* should refer to this Specification for frequency of *Assessment*.

17   COMPLAINTS TO SUPPLIERS

17.1   The *Certification Body* refer to this Specification.

30

Appendix 9 Particular requirements for temporary installations
covering *electrical installation work* for entertainment or events

This Appendix applies to electrical systems as defined in BS 7909 used in structures, sets, mobile units etc as used for public or private events, touring shows, theatrical, radio, TV or film productions and similar activities of the entertainment industry.

1. The requirements set out in this Appendix vary the basic *Criteria* in the body of the Specification.

2. This Appendix sets out the *Criteria* to be used by a body offering accreditation of *Enterprises* involved with temporary power systems predominantly for events and entertainment,

3. For categories of work under section A1.3 the *Enterprise* shall also hold copies of the following:
   - BS 7909: Code of practice for temporary electrical systems for entertainment and related purposes

4. *Enterprises* may also need to hold copies of documents appropriate to the work activity such as the following:
   - Event Safety Guide (HSG195) and web-based successors including the Purple Guide and the HSE portal http://www.hse.gov.uk/event-safety/
   - AEO/AEV/ESSA eGuide
   - National Arenas Association aGuide
   - The Technical Standards for Places of Entertainment (ABTT/IOL/DSA/CIEH)

5. Minimum technical competence *Criteria* required for a *proposed qualified supervisor* for temporary *electrical installation work* for entertainment or events is detailed in Table 4B:

31

Electrical installation competent person schemes

B

B1   Self-certification schemes

Competent persons self-certification schemes under the Building Regulations for Electrical Safety in Dwellings.

NOTE: This list was current at the time of going to press; however, enquirers can obtain an up-to-date list from the Institution of Engineering and Technology website: www.electrical.theiet.org

Electrical installation competent person schemes associated with electrical installations

Type of work	Competent person scheme
Installation of a heating or hot water system, or its associated controls.	Association of Plumbing and Heating Contractors (Certification) Limited, Benchmark Certification Limited, Building Engineering Services Competence Assessment Limited, Certsure LLPHETAS Limited, NAPIT Registration Limited, Oil Firing Technical Association Limited, or Stroma Certification Limited.
Installation of an air conditioning or ventilation system in a dwelling that does not involve work on a system shared with other dwellings.	Building Engineering Services Competence Assessment Limited, Certsure LLP, NAPIT Registration Limited, or Stroma Certification Limited.
Installation of fixed low or extra-low voltage electrical installations in dwellings.	BSI Assurance UK Limited, Benchmark Certification Limited, Building Engineering Services Competence Assessment Limited, Certsure LLP, NAPIT Registration Limited, Oil Firing Technical Association Limited, or Stroma Certification Limited.

Type of work	Competent person scheme
Installation of fixed low or extra-low voltage electrical installations in dwellings, as a necessary adjunct to or arising out of other work being carried out by the registered person.	Association of Plumbing and Heating Contractors (Certification) Limited, Benchmark Certification Limited, Building Engineering Services Competence Assessment Limited, Certsure LLP, NAPIT Registration Limited, or Stroma Certification Limited.
Installation in a building of a system to produce electricity, heat or cooling (a) by microgeneration; or (b) from renewable sources (as defined in Directive 2009/28/EC of the European Parliament and of the Council on the promotion of the use of energy from renewable sources).	Association of Plumbing and Heating Contractors (Certification) Limited, BRE Global Limited, Benchmark Certification Limited, Building Engineering Services Competence Assessment Limited, Certsure LLP, HETAS Limited, NAPIT Registration Limited, Oil Firing Technical Association Limited, or Stroma Certification Limited.

NOTE: Certsure LLP schemes are delivered through the NICEIC and ELECSA, see list below.

B2 Electrical installation third party certification schemes and exemptions from the requirement to give building notice or deposit full plans

Type of work	Third party certification schemes
Electrical installations in dwellings.	NAPIT Registration Limited, or Stroma Certification Limited.

The contact details are below:

APHC ASSOCIATION OF PLUMBING & HEATING CONTRACTORS (CERTIFICATION) LIMITED	Association of plumbing and heating contractors 12 The Pavilions Cranmore Drive Solihull B90 4SB
	Telephone: 0121 711 5030 Fax: 0121 705 7871 E-mail:info@aphc.co.uk
BENCHMARK CERTIFICATION LIMITED CERTIFICATED / UKAS PRODUCT CERTIFICATION 0125C	Benchmark Certification International House, George Curl Way Southampton SO18 2RZ
	Telephone: 02380 517069 E-mail:info@benchmark-cert.co.uk Web:www.benchmark-cert.co.uk
BESCA	Building Engineering Services Competence Assessment Old Mansion House, Eamont Bridge Penrith CA10 2BX
	Telephone: 0800 652 5533 E-mail:info@besca.org.uk Web:www.besca.org.uk
ELECSA NICEIC	Certsure Warwick House Houghton Hall Park Houghton Regis Dunstable LU5 5ZX
	Telephone: 0870 013 0382 Fax: 01582 539090 E-mail:enquiries@certsure.com

HETAS Ltd.
Severn House,
Unit 5, Newtown Trading Estate,
Green Lane, Tewkesbury GL20 8HD

Telephone: 01684 278170
Fax: 01684 273929
E-mail:info@hetas.co.uk

Napit Certification Ltd
4th Floor, Mill 3
Pleasely Vale Business Park
Mansfield
Nottinghamshire NG19 8RL

Telephone: 0845 543 0330
Fax: 0845 543 0332
E-mail:info@napit.org.uk
Web:www.napit.org.uk

OFTEC (Oil Firing Technical Association Ltd)
Foxwood House
Dobbs Lane
Kesgrave
Ipswich IP5 2QQ

Telephone: 0845 658 5080
Fax: 0845 658 5181
E-mail:enquiries@oftec.org
Web:www.oftec.org

Stroma Certification

Telephone: 0845 621 1111
E-mail:info@stroma.com
Web:www.stroma.com

Building Regulations: Third Party Certification Schemes - Conditions of Authorisation

C

Department for Communities and local government requirements

Conditions of Authorisation		Notes on how to demonstrate meeting the conditions
No.	**Section 1 The scheme operator**	
1.	Scheme operator to have a robust and non-discriminatory management, quality and administrative system.	The scheme operator's management, quality and administrative system (including surveillance) may draw upon ISO 9001 for its components. The system shall be documented.
2.	Scheme operator to appoint an independent third party acceptable to DCLG to periodically audit their performance against these conditions of authorisation.	Before appointing an independent body to undertake the audit, the scheme operator shall consult DCLG on their suitability and terms of reference. The audit shall be undertaken annually for the first two years. Thereafter a programme of audit should be submitted to DCLG for approval.

C

Conditions of Authorisation	Notes on how to demonstrate meeting the conditions
3. Scheme operator, including assessors and inspectors that it employs, to have the technical ability to assess/inspect the competence of prospective and existing members to certify compliance with the relevant requirements of the Building Regulations.	The scheme operator shall document how the assessment process will be undertaken and what level of competence each assessor and inspector has, which must be at least to the equivalent level required of members in the Minimum Technical Competence (MTC) assessment procedure (see condition no 9).
4. Scheme operator to ensure that the scheme is financially viable and self-sufficient within a reasonable timescale.	The scheme operator shall: (a) have a transparent fee structure showing income from members and how the scheme will be self-financing with a sufficient surplus for development; (b) ensure that the scheme is self-financing, within a period of not later than five years after authorisation; (c) use scheme funds received from members from registration and notification fees etc. only for the benefit of members of the scheme. This can include use of funds for the general benefit of the sector in which the scheme operates.
5. Scheme operator to have an absence of, or methods for avoiding, conflicts of interest between the commercial interests of any sponsoring or parent organisations and management of the scheme.	The scheme operator shall ensure that certifiers acting under their scheme do not certify electrical work undertaken by employees of sponsoring or parent organisations. The scheme operator shall document how any conflicts of interest will be managed. For example: possible conflicts of interest may arise where a scheme is part of or owned by a larger commercial, trade or professional body.
6. Scheme operator to provide annual accounts, independently audited, for the scheme itself.	This condition will support conditions 4 and 5 as the accounts will help show that a scheme is financially viable and self-sufficient. It will also help demonstrate that there is not a financial conflict of interest. To be 'independently audited', accounts must have been checked by someone who is competent to check them and who is independent of the preparation of the accounts.

Conditions of Authorisation	Notes on how to demonstrate meeting the conditions
Section 2 The scheme operator and its members	
7. Scheme operator to establish and publish scheme rules, including its application and certification processes and fee structure.	The scheme rules shall be published on the scheme operator's public website as a minimum.
8. Scheme operator to assess applicants as technically competent, against National Occupational Standards (NOS) under the relevant Minimum Technical Competence (MTC) assessment procedure, before registering them with the scheme. The assessment must include witnessing individual certifiers carrying out on-site inspection and testing and completing appropriate documentation.	The relevant MTC assessment procedure must be used. The scheme rules shall set out details of how the technical competence of applicants will be assessed.
9. Scheme operator to ensure that its members' competencies are kept up to date (for example as a result of changes to the Building Regulations and/or BS EN standards or technical approvals).	This may be by means of formal generic training courses, seminars, distance learning, etc. as appropriate, which shall be equally available to all its members.
10. Scheme operator to issue photo ID cards to the certifiers listing the competence of the certifier concerned.	The ID card should include the following as a minimum: photo & name of certifier, name of company, name of scheme, registration number with scheme, contact point for queries.
11. Scheme operator to ensure that all registrants have a copy of the current edition of the IET *Electrician's Guide to the Building Regulations*, and to provide ongoing technical and other help and advice as required post registration.	Provision of ongoing advice may be accomplished by setting up telephone and e-mail helplines available to all members.
12. Scheme operator to undertake surveillance of its members' work, including carrying out periodic random checking of a representative sample of each certifier's inspections, to check compliance with the Building Regulations 2010.	Inspections will be undertaken annually for the first two years following initial assessment. Thereafter, a scheme operator shall, under a risk-based approach, undertake a minimum of one on-site witnessed assessment every three years for each individual certifier with a clean track record, as defined and documented by the scheme operator. The decision on whether or not to inspect more frequently will need to be based on factors such as inspection outcomes, significant complaints and changes of certifiers within a business. The scheme operator shall document its surveillance process and keep records of all surveillance activity undertaken.

Conditions of Authorisation	Notes on how to demonstrate meeting the conditions
13. Scheme operator to have effective sanctions in place for dealing with scheme members and/or certifiers who fail to take all reasonable steps to determine non-compliance with the Building Regulations and/or a breach of scheme rules by members of the scheme.	The scheme rules shall set out the range of sanctions to be applied in particular circumstances, including requiring the certifier to re-do their inspection and testing at no additional cost to the customer; referral to the local authority if a certificate is incorrect; and, in the last resort, termination of membership for refusal or inability to comply. Provision shall also be made for an appeal against any sanctions imposed.
14. Scheme operator to use an agreed mechanism to make available to other scheme operators and other interested parties (e.g. LABC & relevant Government Departments) the names of former members and individual certifiers whose membership has been terminated by the scheme and the reason for termination.	This applies where the reasons for termination of membership relate to non-compliance with the Building Regulations or a breach of scheme rules. The names of such former members shall remain available for a period of at least two years. All members and individual certifiers must be made aware of this condition on initial registration and/or renewal of membership.
Section 3 The scheme operator and its customers	
15. Scheme operator to publicise the existence of its scheme and to keep and publish membership lists and lists of individual certifiers who have been assessed as competent.	This information shall be published on the scheme operator's public website as a minimum. Publication is subject to the consent of members and individual certifiers, as a condition of membership, and must be sought from members and certifiers on initial registration and/or renewal of membership.

Conditions of Authorisation	Notes on how to demonstrate meeting the conditions
16. Scheme operator to have a robust and publicised complaints procedure.	The complaints procedure must cover those relating to the activity of checking for non-compliance with the Building Regulations, but may include other types of complaints from customers (and members) relating to the scheme (e.g. complaints relating to negligence, incompetence or dishonesty on the part of the certifier). It should also cover complaints against the scheme operator. The stages of the scheme's complaints procedure shall be set out in detail, at a minimum on its public website, so that those wishing to use the procedure are aware of the stages. The procedures shall be consistent with the principles relating to complaints management of the Trading Standards Institute Consumer Codes Approval Scheme.
17. Scheme operator to require members to carry professional indemnity insurance.	Members must have professional indemnity insurance to cover liabilities up to the level prescribed in the relevant MTC.
18. Scheme operator to require its members to remain responsible for ensuring that all certification work is carried out under a contract with the customer and is compliant with the Building Regulations.	This shall be stated in the scheme rules. The scheme rules shall also state that work should only be undertaken by a certifier registered by the scheme member.
19. Scheme operator to take measures to ensure that it is notified by members of all certified work required under the scheme and to forward to all customers a certificate of Building Regulations compliance.	The scheme operator shall have documented systems in place to ensure that members are notifying all jobs certified under the scheme, in line with regulation 20A of the Building Regulations 2010 as amended. The scheme operator should receive notifications well within time to ensure that it meets the 30 calendar day deadline for giving compliance certificates to customers.

C

Conditions of Authorisation	Notes on how to demonstrate meeting the conditions
Section 4 The scheme operator and DCLG/local authorities	
20. Scheme operator to provide the information DCLG requires in order to carry out its oversight functions, both on a regular basis or *ad hoc* as required.	DCLG will specify with all scheme operators the regular information needed for its purposes, which may be published on its website. This is likely to include: • provision of a periodic report on: membership numbers; number of notifications made to local authorities; and number of formal consumer complaints and their outcomes; • financial information annually from the scheme's audited accounts, as appropriate; and • information from the appointed auditors (DCLG may request this direct from the auditors).
21. Scheme operator to take measures to ensure that it is notified by members of all certified work required under the scheme and to forward this information to the relevant local authority in the format agreed with LABC.	As under condition 19, scheme operators shall have documented systems in place to ensure that members are notifying all jobs certified under the scheme, in line with regulation 20A of the Building Regulations 2010. The scheme operator should receive notifications well within time to ensure it meets the 30 calendar day deadline for transfer of information to the local authority.

Department for Communities and Local Government October 2013

Approved Document B: Fire safety, Volume 1 – Dwellinghouses

D

Extracts from Approved Document B, Volume 1.

NOTE: Approved document B1 2006 edition incorporating 2010 and 2013 amendments refers to BS 5839-1:2002 and BS 5839-6:2004. Whilst these are not the latest versions of the standards, the clauses relevant to the text of the approved document are unchanged.

Section 1: Fire detection and fire alarm systems

Introduction

1.1 Provisions are made in this section for suitable arrangements to be made in dwelling-houses to give early warning in the event of fire.

General

1.2 The installation of smoke alarms, or automatic fire detection and alarm systems can significantly increase the level of safety by automatically giving an early warning of fire. The following guidance is appropriate for most dwellinghouses. However, where it is known that the occupants of a proposed dwellinghouse are at a special risk from fire, it may be more appropriate to provide a higher standard of protection, e.g. additional detectors.

1.3 All new dwellinghouses should be provided with a fire detection and fire alarm system in accordance with the relevant recommendations of BS 5839-6:2004 to at least a Grade D Category LD3 standard.

1.4 The smoke and heat alarms should be mains-operated and conform to BS 54461:2000 or BS 5446-2:2003, respectively: Fire detection and fire alarm devices

D

for dwelling-houses. Part 1 Specification for smoke alarms: or Part 2 Specification for heat alarms. They should have a standby power supply, such as a battery (either rechargeable or non-rechargeable) or capacitor. More information on power supplies is given in clause 15 of BS 5839-6:2004.

NOTE: BS 5446-1 covers smoke alarms based on ionization chamber smoke detectors and optical (photoelectric) smoke detectors. The different types of detector respond differently to smouldering and fast-flaming fires. Either type of detector is generally suitable. However, the choice of detector type should, if possible, take into account the type of fire that might be expected and the need to avoid false alarms. Optical detectors tend to be less affected by low levels of 'Invisible' particles, such as fumes from kitchens, that often cause false alarms. Accordingly, they are generally more suitable than ionization chamber detectors for installation in circulation spaces adjacent to kitchens.

Large houses

1.5 A dwellinghouse is regarded as large if it has more than one storey and any of those storeys exceed 200 m².

1.6 A large dwellinghouse of 2 storeys (excluding basement storeys) should be fitted with a fire detection and fire alarm system of Grade B Category LD3 as described in BS 58396:2004.

1.7 A large dwellinghouse of 3 or more storeys (excluding basement storeys) should be fitted with a Grade A Category LD2 system as described in BS 5839-6:2004, with detectors sited in accordance with the recommendations of BS 5839-1:2002 for a Category L2 system.

Material alterations

1.8 Where new habitable rooms are provided above the ground floor level, or where they are provided at ground floor level and there is no final exit from the new room, a fire detection and fire alarm system should be installed. Smoke alarms should be provided in the circulation spaces of the dwellinghouse in accordance with paragraphs 1.10 to 1.18 to ensure that any occupants of the new rooms are warned of any fire that may impede their escape.

Sheltered housing

1.9 The detection equipment in a sheltered housing scheme with a warden or supervisor should have a connection to a central monitoring point (or alarm receiving centre) so that the person in charge is aware that a fire has been detected in one of the dwellinghouses and can identify the dwellinghouse concerned. These provisions are not intended to be applied to the common parts of a sheltered housing development, such as communal lounges, or to sheltered accommodation in the Institutional or Other residential purpose groups (see Approved Document B, Volume 2).

Positioning of smoke and heat alarms

1.10 Detailed guidance on the design and installation of fire detection and alarm systems in dwellinghouses is given in BS 5839-6:2004. However, the following guidance is appropriate to most common situations.

1.11 Smoke alarms should normally be positioned in the circulation spaces between sleeping spaces and places where fires are most likely to start (e.g. kitchens and living rooms) to pick up smoke in the early stages of a fire.

1.12 There should be at least one smoke alarm on every storey of a dwellinghouse.

1.13 Where the kitchen area is not separated from the stairway or circulation space by a door, there should be a compatible interlinked heat detector or heat alarm in the kitchen, in addition to whatever smoke alarms are needed in the circulation space(s).

1.14 Where more than one alarm is installed they should be linked so that the detection of smoke or heat by one unit operates the alarm signal in all of them. The manufacturers' instructions about the maximum number of units that can be linked should be observed.

1.15 Smoke alarms/detectors should be sited so that:

(a) there is a smoke alarm in the circulation space within 7.5 m of the door to every habitable room;

(b) they are ceiling-mounted and at least 300 mm from walls and light fittings (unless, in the case of light fittings, there is test evidence to prove that the proximity of the light fitting will not adversely affect the efficiency of the detector). Units designed for wall-mounting may also be used provided that the units are above the level of doorways opening into the space and they are fixed in accordance with manufacturers' instructions; and

(c) the sensor in ceiling-mounted devices is between 25 mm and 600 mm below the ceiling (25–150 mm in the case of heat detectors or heat alarms).

NOTE: This guidance applies to ceilings that are predominantly flat and horizontal.

1.16 It should be possible to reach the smoke alarms to carry out routine maintenance, such as testing and cleaning, easily and safely. For this reason, smoke alarms should not be fixed over a stair or any other opening between floors.

1.17 Smoke alarms should not be fixed next to or directly above heaters or air-conditioning outlets. They should not be fixed in bathrooms, showers, cooking areas or garages, or any other place where steam, condensation or fumes could give false alarms.

1.18 Smoke alarms should not be fitted in places that get very hot (such as a boiler room) or very cold (such as an unheated porch). They should not be fixed to surfaces that are normally much warmer or colder than the rest of the space, because the temperature difference might create air currents which move smoke away from the unit.

D

Power supplies

1.19 The power supply for a smoke alarm system should be derived from the dwelling house's mains electricity supply. The mains supply to the smoke alarm(s) should comprise a single independent circuit at the dwellinghouse's main distribution board (consumer unit) or a single regularly used local lighting circuit. This has the advantage that the circuit is unlikely to be disconnected for any prolonged period. There should be a means of isolating power to the smoke alarms without isolating the lighting.

1.20 The electrical installation should comply with Approved Document P: Electrical safety – Dwellings.

1.21 Any cable suitable for domestic wiring may be used for the power supply and interconnection to smoke alarm systems. It does not need any particular fire survival properties except in large houses (BS 5839-6:2004 specifies fire resisting cables for Grade A and B systems). Any conductors used for interconnecting alarms (signalling) should be readily distinguishable from those supplying mains power, e.g. by colour coding.

NOTE: Mains-powered smoke alarms may be interconnected using radio-links, provided that this does not reduce the lifetime or duration of any standby power supply below 72 hours. In this case, the smoke alarms may be connected to separate power circuits (see paragraph 1.19)

1.22 Other effective options exist and are described in BS 5839-1:2002 and BS 58396:2004. For example, the mains supply may be reduced to extra-low voltage in a control unit incorporating a standby trickle-charged battery, before being distributed at that voltage to the alarms.

Design and installation of systems

1.23 It is essential that fire detection and fire alarm systems are properly designed, installed and maintained. Where a fire alarm system is installed, an installation and commissioning certificate should be provided. Third-party certification schemes for fire protection products and related services are an effective means of providing the fullest possible assurances, offering a level of quality, reliability and safety.

1.24 A requirement for maintenance cannot be made as a condition of passing plans by the Building Control Body. However, the attention of developers and builders is drawn to the importance of providing the occupants with information on the use of the equipment, and on its maintenance (or guidance on suitable maintenance contractors). See paragraph 0.11.

NOTE: BS 5839-1 and BS 5839-6 recommend that occupiers should receive the manufacturers' instructions concerning the operation and maintenance of the alarm system.

Section 5: Compartmentation

Introduction

5.1 The spread of fire within a building can be restricted by sub-dividing it into compartments separated from one another by walls and/or floors of fire-resisting construction. The object is twofold:

(a) to prevent rapid fire spread, which could trap occupants of the building; and

(b) to reduce the chance of fires becoming large, on the basis that large fires are more dangerous, not only to occupants and fire and rescue service personnel, but also to people in the vicinity of the building. Compartmentation is complementary to provisions made in Section 2 for the protection of escape routes, and to provisions made in Sections 8 to 10 against the spread of fire between buildings.

Provision of compartmentation

5.2 Compartment walls and compartment floors should be provided in the circumstances described below, with the proviso that the lowest floor in a building does not need to be constructed as a compartment floor. Provisions for the protection of openings in compartment walls and compartment floors are given in paragraph 5.13 and Section 7.

Houses

5.3 Every wall separating semi-detached houses, or houses in terraces, should be constructed as a compartment wall, and the houses should be considered as separate buildings.

5.4 If a domestic garage is attached to (or forms an integral part of) a house, the garage should be separated from the rest of the house, as shown in Diagram 10.

5.5 Where a door is provided between a dwellinghouse and the garage, the floor of the garage should be laid to allow fuel spills to flow away from the door to the outside. Alternatively the door opening should be positioned at least 100 mm above floor level.

▼ **Diagram 10**   Separation between garage and dwellinghouse

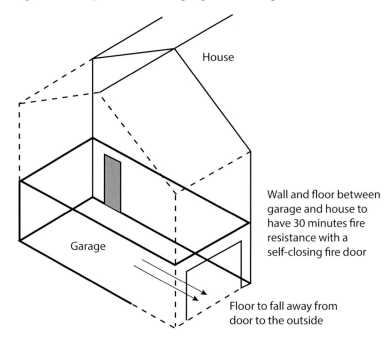

House

Garage

Wall and floor between
garage and house to
have 30 minutes fire
resistance with a
self-closing fire door

Floor to fall away from
door to the outside

Flats (from vol. 2)

8.13 In buildings containing flats, the following should be constructed as compartment walls or compartment floors:

- **(a)** every floor (unless it is within a flat, i.e. between one storey and another within one individual dwelling); and
- **(b)** every wall separating an apartment from any other part of the building; and (**NOTE:** 'any other part of the building' does not include an external balcony/deck access)
- **(c)** every wall enclosing a refuse storage chamber.

Section 7: Protection of openings and fire-stopping

Introduction

7.1 Sections 7 and 8 make provisions for fire-separating elements, and set out the circumstances in which there may be openings in them. This section deals with the protection of openings in such elements.

7.2 If a fire-separating element is to be effective, then every joint, or imperfection of fit, or opening to allow services to pass through the element, should be adequately protected by sealing or fire-stopping so that the fire resistance of the element is not impaired.

7.3 The measures in this section are intended to delay the passage of fire. They generally have the additional benefit of retarding smoke spread, but the test specified in Appendix A of Approved Document B for integrity does not stipulate criteria for the passage of smoke.

7.4 Consideration should be given to the effect of services that may be built in to the construction that could adversely affect its fire resistance. For instance when downlighters, loudspeakers and other electrical accessories are installed, additional protection may be required to maintain the integrity of a wall or floor.

7.5 Detailed guidance on door openings and fire doors is given in Appendix B of Approved Document B.

▼ **Table 3** Maximum nominal internal diameter of pipes passing through a fire separating element (see paragraphs 7.6 to 7.9)

Situation	Pipe material and maximum nominal internal diameter (mm)		
	(a) Non-combustible material[1]	(b) Lead, aluminium, aluminium alloy, uPVC[2], fibre cement	(c) Any other material
1. Wall separating dwellinghouses	160	160 (stack pipe)[3] 110 (branch pipe)[3]	40
2. Wall or floor separating a dwellinghouse from an attached garage	160	110	40
3. Any other situation	160	40	40

NOTES:

1. Any non-combustible material (such as cast iron, copper or steel) which, if exposed to a temperature of 800 °C, will not soften or fracture to the extent that flame or hot gas will pass through the wall of the pipe.
2. uPVC pipes complying with BS 4514 and uPVC pipes complying with BS 5255.
3. These diameters are only in relation to pipes forming part of an above-ground drainage system and enclosed as shown in Diagram 15. In other cases the maximum diameters against situation 3 apply.

Openings for pipes

7.6 Pipes that pass through fire-separating elements (unless the pipe is in a protected shaft), should meet the appropriate provisions in alternatives A, B or C below.

Alternative A: Proprietary seals (any pipe diameter)

7.7 Provide a proprietary sealing system which has been shown by test to maintain the fire resistance of the wall, floor or cavity barrier.

Alternative B: Pipes with a restricted diameter

7.8 Where a proprietary sealing system is not used, fire-stopping may be used around the pipe, keeping the opening as small as possible. The nominal internal diameter of the pipe should not be more than the relevant dimension given in Table 3.

Alternative C: Sleeving

7.9 A pipe of lead, aluminium, aluminium alloy, fibre-cement or uPVC, with a maximum nominal internal diameter of 160 mm, may be used with a sleeving of non-combustible pipe as shown in Diagram 14. The specification for non-combustible and uPVC pipes is given in the notes to Table 3.

▼ **Diagram 14** Pipes penetrating structure (see paragraph 7.9)

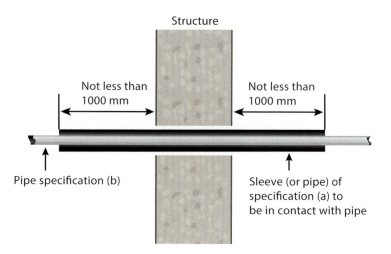

NOTES:

(a) Make the opening in the structure as small as possible and provide fire-stopping between pipe and structure.

(b) See Table 3 for materials specification.

Safety instructions

E

Appendix C is a typical set of company safety instructions that could be used as the basis for a particular organisation.

Part 1: General provisions

Scope

These Safety Instructions are for general application for work involving either, or both, non-electrical and electrical work as further described in Parts 2 and 3.

Definitions

(For use with this safety instruction.)

Appliance

A device requiring a supply of electricity to make it work.

Approved

Sanctioned in writing by the responsible director in order to satisfy in a specified manner the requirements of any or all of these Safety Instructions.

Company

To be defined.

Competent person

Person required to work on electrical equipment, installations and appliances and recognised by the Employer as having sufficient technical knowledge and/or experience to enable him/her to carry out the specified work properly without danger to themselves or others. It is recommended that this competence should be recognised by means of written documentation.

Conductor

An electrical conductor arranged to be electrically connected to a system.

Customer

A person, or body, that has a contractual relationship with the Employer for the provision of goods or services.

Danger

Risk of injury to persons (and livestock where expected to be present) from:

(a) fire, electric shock and burns arising from the use of electrical energy; and

(b) mechanical movement of electrically controlled equipment, insofar as such danger is intended to be prevented by electrical emergency switching or by electrical switching for mechanical maintenance of non-electrical parts of such equipment.

Dead

At or about zero voltage in relation to earth, and disconnected from any live system.

Earth

The conductive mass of the Earth, whose electric potential at any point is conventionally taken as zero.

Earthed

Connected to earth through switchgear with an adequately rated earthing capacity or by approved earthing leads.

Electrical equipment

Anything used, intended to be used or installed for use to generate, provide, transmit, transform, rectify, convert, conduct, distribute, control, store, measure or use electrical energy.

Electrical installation

An assembly of associated electrical equipment supplied from a common origin to fulfil a specific purpose and having certain coordinated characteristics.

Isolated

The disconnection and separation of the electrical equipment from every source of electrical energy in such a way that this disconnection and separation is secure.

Live

Electrically charged.

Notices

Caution Notice – A notice in approved form conveying a warning against interference.

Danger Notice – A notice in approved form reading 'Danger'.

Responsible Director

The Director of the Company, partner or owner responsible for safety.

Supervisor

(a) Immediate Supervisor – a person (having adequate technical knowledge, experience and competence) who is regularly available at the location where work is in progress or who attends the work area as is necessary to ensure the safe performance and completion of work.

(b) Personal Supervisor – a person (having adequate technical knowledge, experience and competence) such that he/she is at all times during the course of the work in the presence of the person being supervised.

Voltage

Voltage by which an installation (or part of an installation) is designated. The following ranges of nominal voltage (rms values for a.c.) are defined:

▶ Extra-low. Normally not exceeding 50 V a.c. or 120 V ripple-free d.c., whether between conductors or to Earth.

▶ Low. Normally exceeding extra-low but not exceeding 1000 V a.c. or 1500 V d.c. between conductors, or 600 V a.c. or 900 V d.c. between conductors and Earth.

The actual voltage of the installation may differ from the nominal value by a quantity within normal tolerances.

▶ High. Normally exceeding 1000 V a.c. or 1500 V d.c. between conductors, or 600 V a.c. or 900 V d.c. between conductors and Earth.

Basic requirements

E1.1 Other safety rules and related procedures

In addition to the application of these Safety Instructions, other rules and procedures as issued by the Employer, or by other authorities, shall be complied with in accordance with management instructions. In that employees may be required to work in locations, or on or near electrical equipment, installations and appliances, that are not owned or controlled by the Employer, these Safety Instructions have been produced to reasonably ensure safe working, since no other rules/instructions will normally be applicable. However, where the owner has his own rules/instructions and procedures, agreement shall be reached between the Company and the owner on which rules/instructions shall be applied. Such agreement shall be made known to the employees concerned.

E1.2 Information and instruction

Arrangements shall be made to ensure:

(a) that all employees concerned are adequately informed and instructed as to any equipment, installations or appliances which are associated with work and which legal requirements, Safety Rules and related procedures shall apply; and

(b) that other persons who are not employees but who may be exposed to danger by the work also receive reasonably adequate information.

E1.3 Issue of Safety Instructions

Employees and other persons issued with safety instructions shall sign a receipt for a copy of these Safety Instructions (and any amendments thereto) and shall keep them in good condition and have them available for reference as necessary when work is being carried out under these Safety Instructions.

E1.4 Special procedures

Work on, or test of, equipment, installations and appliances to which rules cannot be applied, or for special reasons should not be applied, shall be carried out in accordance with recognised good practice.

E1.5 Objections

When any person receives instructions regarding work covered by these Safety Instructions and objects, on safety grounds, to the carrying out of such instructions, the person issuing them shall have the matter investigated and, if necessary, referred to a higher authority for a decision before proceeding.

E1.6 Reporting of accidents and dangerous occurrences

All accidents and dangerous occurrences, whether of an electrical nature or not, shall be reported in accordance with the Reporting of Injuries, Diseases and Dangerous Occurrences Regulations 1995.

E1.7 Health and safety

The employer and all employees have a duty to comply with the relevant provisions of the Health and Safety at Work etc. Act 1974 and with other relevant statutory provisions and the various Regulations affecting health and safety, including electrical safety. Additionally, authoritative guidance is available from the Health and Safety Executive and other sources. In addition to these statutory duties and any other responsibilities separately allocated to them, all persons who may be concerned with work as detailed in Section B1.1 shall be conversant with, and comply with, those Safety Instructions and codes of practice relevant to their duties. Ignorance of legal requirements, or of Safety Instructions and related procedures shall not be accepted as an excuse for neglect of duty. If any person has any doubt as to any of these duties he should report the matter to his immediate supervisor.

E1.8 Compliance with Safety Instructions

It is the duty of everyone who may be concerned with work covered by these Safety Instructions, to ensure their implementation and to comply with them and related codes of practice. Ignorance of the relevant legal requirements, Safety Instructions, Codes of Practice or approved procedures is not an acceptable excuse for neglect of duty.

The responsibilities placed upon persons may include all or part of those detailed in this section, depending on the role of the persons. Any written authorisation given to persons to perform their designated role in implementing the Safety Instructions must indicate the work permitted. Whether employees are authorised as competent or not, all have the following duties that they must ensure are implemented:

- ▶ All employees shall comply with these Safety Instructions when carrying out work, whether instructions are issued orally or in writing.
- ▶ All employees shall use safe methods of work, safe means of access and the personal protective equipment and clothing provided for their safety.
- ▶ All employees when in receipt of work instructions shall:

  (a) be fully conversant with the nature and extent of the work to be done;

  (b) read the contents and confirm to the person issuing the instructions that they are fully understood;

  (c) during the course of the work, adhere to, and instruct others under their charge to adhere to, any conditions, instructions or limits specified in the work instructions; and

  (d) when in charge of work, provide immediate or personal supervision as required.

Part 2: Non-electrical

Scope

The non-electrical part of the Safety Instructions shall be applied to work by employees in the activities that are non-electrical. This work may involve:

(a) work on customers' premises;

(b) work on employers' premises;

(c) work on the public highway or in other public places.

The Safety Instructions applicable to this work are those contained in Parts 1 and 2 of the Safety Instructions. When work of an electrical nature is being carried out, all Parts (1, 2 and 3) of the Safety Instructions apply.

Basic safety precautions

E2.1 General principle

The general principle is to avoid accidents. Most accidents arise from simple causes and can be prevented by taking care.

E2.2 Protective clothing and equipment

The wearing of protective clothing and the use of protective equipment can, in appropriate circumstances, considerably reduce the severity of injury should an accident occur. Where any work under these Safety Instructions and related procedures takes place, appropriate safety equipment and protective clothing of an approved type shall be issued and used. At all times employees are expected to wear sensible clothing and footwear having regard to the work being carried out. Further references are made, in particular circumstances, to the use of gloves. Hard hats must be worn at all times when there is a risk of head injury and particularly on building sites. Where there is danger from flying particles of metal, concrete or stone, suitable eye protection must be provided and must be used by employees. If necessary, additional screens must be provided to protect other persons in the vicinity.

E2.3 Good housekeeping

Tidiness, wherever work is carried out, is the foundation of safety; good housekeeping will help to ensure a clean, tidy and safe place of work.

Particular attention should be paid to:

(a) picking up dropped articles immediately;
(b) wiping up any patches of oil, grease or water as soon as they appear and, if necessary, spreading sand or sawdust;
(c) removing rubbish and scrap to the appropriate place;
(d) preventing objects falling from a height by using containers for hand tools and other loose material;
(e) ensuring stairs and exits are kept clear.

No job is completed until all loose gear, tools and materials have been cleared away and the workplace left clean and tidy. Most falls are caused by slippery substances or loose objects on the floor and good housekeeping will remove most of the hazards that can occur.

E2.4 Safe access

It is essential that every place of work is at all times provided with safe means of access and exit, and these routes must be maintained in a safe condition. Keeping the workplace tidy minimises the risk of falling which is the major cause of accidents, but certain special hazards associated with work in confined spaces require particular attention.

E2.4.1 Ladders

All ladders should be of sound construction, uniquely identified and free from apparent defects. This is of particular importance in connection with timber ladders. The following practices should always be observed:

Ladders should be checked before use. Any defects must be reported and the equipment clearly marked and not used until repaired. All ladders should be regularly inspected by a competent person and a record kept. Ladders in use should stand on a level and firm footing. Loose packing should not be used to support the base. Ladders should be used at the correct angle, i.e., for every four metres up, the bottom of the ladder must be one metre out. Ladders should be lashed at the top when in use, but when this is not practicable they should be held secure at the bottom. The ladder top should extend to a height of at least one metre above any landing place. Hand tools and other material should not be carried in the hand when ascending or descending ladders. A bag and sash line should be used. Suitable crawling ladders or boards must be used when working on asbestos, cement and other fragile roofs. Permanent warning notices should be placed at the means of access to these roofs. (NB: In a situation where no ladder is available, and the work requires a small step up, it is the employee's responsibility to ensure that any other article used for the purpose is totally suitable.)

E2.4.2 Openings in floors

Every floor opening must be guarded, and it is important that other occupants of the workplace are made aware of these hazards. In addition to the risk of persons falling through any opening, there is also a risk from falling objects, and safe placing of tools, materials and other objects when working near openings, holes or edges, or at any height, will also prevent accidents. If work has to be carried out in confined spaces such as tunnels and underground chambers, the atmosphere may be deficient in oxygen or may contain dangerous fumes or substances. The Energy Networks Association Engineering Recommendation, ERG64, Safety in Cableways or similar must be followed in these circumstances.

E2.5 Lighting

Good lighting, whether natural or artificial, is essential to the safety of people whether at the workplace or moving about. If natural lighting is inadequate, it must be supplemented by adequate and suitable artificial lighting. If danger may arise from a power failure, adequate emergency lighting is required.

E2.6 Lifting and handling

All employees must be trained in the appropriate lifting and handling techniques according to the type of work undertaken.

E2.7 Fire precautions

All employees must be thoroughly conversant with the procedure to be followed in the event of fire. Whether working on customers' premises or elsewhere, employees should familiarise themselves with escape routes, fire precautions, etc., before commencing work. Fire exits must always be kept clear, and access to fire fighting equipment unobstructed. All fire fighting equipment that is the Employer's responsibility must be regularly inspected, maintained and recorded whether by local supervisor, safety supervisor or appropriate third party. Individuals should report any apparent damage to equipment.

E2.8 Hand tools

Hand tools must be suitable for the purpose for which they are being used and are the responsibility of those using them. They must be maintained in good order and any which are worn or otherwise defective must be reported to a supervisor. Approved insulated tools must be available for work on live electrical equipment.

E2.9 Mechanical handling

Fork-lift and similar trucks must only be driven by operators who have been properly trained, tested and certified for the type of trucks they have to operate. Supervisors should control the issue and return of the truck keys and they should ensure that a daily check of the truck and its controls is carried out by the operator.

E2.10 Portable power tools

All portable electrical apparatus including cables, portable transformers and other ancillary equipment should be inspected before use and maintained and tested at regular intervals.

Trailing cables are frequently damaged and exposed to wet conditions. Users must report all such damage and other defects as soon as possible, and the faulty equipment must be immediately withdrawn from use. When not in use, power tools should be switched off and disconnected from the source of supply.

E2.11 Welding, burning and heating processes

E2.11.1   General

Welding, burning and heating processes involving the use of gas and electricity demand a high degree of skill and detailed knowledge of the appropriate safety requirements. Specific safety instructions will be issued to employees using such equipment. Suitable precautions should be taken, particularly when working overhead, to prevent fire or other injury from falling or flying sparks. All heating, burning and welding equipment must be regularly inspected, and a record kept.

E2.11.2   Propane

Propane is a liquefied petroleum gas stored under pressure in cylinders which must be stored vertically in cool, well-ventilated areas, away from combustible material, heat sources and corrosive conditions. Cylinders must be handled carefully and not allowed to fall from a height; when transported, they must always be carried in an approved restraint. When the cylinder valve is opened the liquid boils, giving off a highly flammable gas. The gas is heavier than air and can give rise to a highly explosive mixture. It is essential, therefore, that valves are turned off after use.

E2.12 Machinery

All machinery shall be guarded as necessary to prevent mechanical hazards. Facilities shall be provided for isolating and locking off the power to machinery. Work on machines shall not commence unless isolation and locking off from all sources of power has been effected and permits to work issued.

Part 3: Electrical

Scope

This part of the Safety Instructions shall be applied to electrical work. This work may involve:

(a) employers' equipment;
(b) customers' electrical installations; and
(c) customers' electrical appliances.

This work will normally be concerned with equipment, installations and appliances at low voltages. In the event of work needing to be carried out on high voltage equipment and installations (i.e. where the voltage exceeds 1000 volts a.c.), additional instructions and procedures laid down for high voltage work must be issued to those employees who carry out this work.

Basic safety precautions

E3.1 General principle

As a general principle, and wherever reasonably practicable, work should only be carried out on equipment that is dead and isolated from all sources of supply. Such equipment should be proved dead by means of an approved voltage testing device which should be tested before and after verification, or by clear evidence of isolation taking account of the possibility of wrong identification or circuit labelling. Equipment should always be assumed to be live until it is proved dead. This is particularly important where there is a possibility of backfeed from another source of supply.

E3.2 Information prior to commencement of work

According to the complexity of the installation, the following information may need to be provided before specified work is carried out:

(a) details of the supply to the premises, and to the system and equipment on which work is to be carried out;
(b) details of the relevant circuits and equipment and the means of isolation;
(c) details of any customer's safety rules or procedures that may be applicable to the work;
(d) the nature of any processes or substances which could give rise to a hazard associated with the work, or other special conditions that could affect the working area, such as the need for special access arrangements;
(e) emergency arrangements on site; and
(f) the name and designation of the person nominated to ensure effective liaison during the course of the work.

Where the available information, or the action to be taken as a result of it, is considered by the person in charge of the work to be inadequate for safe working, such work should not proceed until that inadequacy has been removed or a decision obtained from a person in higher authority. Defects affecting safe working should be reported to the appropriate supervisor.

E3.3 Precautions to be taken before work commences on dead electrical equipment

In addition to any special precautions to be taken at the site of the work, such as for the presence of hazardous processes or substances, the following electrical precautions should be taken, according to the circumstances, before work commences on dead electrical equipment.

(a) The electrical equipment should first be properly identified and disconnected from all points of supply by the opening of circuit-breakers, isolating switches, the removal of fuses, links or current-limiting devices, or other suitable means. Approved Notices, warning against interference, should be affixed at all points of disconnection.

(b) All reasonably practicable steps should be taken to prevent the electrical equipment being made live inadvertently. This may be achieved, according to the circumstances, by taking one or more of the following precautions:

   (i) approved locks should be used to lock off all switches etc. at points where the electrical equipment and associated circuits can be made live. This should be additional to any lock applied by any other party; the keys to all locks should be retained by the person in charge of the work or in a specially provided key safe;

   (ii) any fuses, links or current-limiting devices involved in the isolation procedures should be retained in the possession of the person in charge of the work;

   (iii) in the case of portable apparatus, where isolation has been by removal of a plug from a socket-outlet, suitable arrangements should be made to prevent unauthorised re-connection; and

   (iv) approved notices should be placed at points where the electrical equipment and associated circuits can be made live.

(c) The electrical equipment should be proved dead by the proper use of an approved voltage testing device and/or by clear evidence of isolation, such as physically tracing a circuit. Approved testing devices should be checked immediately before and after use to ensure that they are in working order.

(d) When work is carried out on timeswitched or other automatically controlled equipment or circuits, the fuses or other means of isolation controlling such equipment or circuits should be removed. On no account should reliance be placed on the timeswitches, limit switches, lock-out push buttons etc., or on any other auxiliary equipment, as means of isolation.

(e) Where necessary, approved notices should be displayed to indicate any exposed live conductors in the working zone.

(f) When it is required to work on dead equipment situated in a substation or similar place where there are exposed live conductors, or adjacent to high voltage plant, the safe working area should be defined by a person authorised in writing under the Safety Rules or under procedures controlling that plant, and all subsequent work must be conducted in accordance with such rules or procedures. Where necessary, the exposed live conductors should be adequately screened in an approved manner or by other approved means taken to avoid danger from the live conductors.

E3.4 Precautions to be taken before work commences on or near live equipment

No person shall be engaged in any work activity on or so near any live conductor (other than one suitably covered with insulating material so as to prevent danger) that danger may arise unless:

(a) it is unreasonable in all the circumstances for it to be dead; and

(b) it is reasonable in all the circumstances for them to be at work on or near it while it is live; and

(c) suitable precautions (including where necessary the provision of suitable protective equipment) are taken to prevent injury (regulation 14, Electricity at Work Regulations). Where work is to be carried out on live equipment the following protective equipment should be provided, maintained and used, by adequately trained personnel, in accordance with the Safety Rules or local procedures as appropriate:

(i) approved screens or screening material;

(ii) approved insulating stands in the form of hardwood gratings or approved rubber insulating mats;

(iii) approved insulated tools; and

(iv) approved insulating gloves.

When testing, including functional testing or adjustment of electrical equipment, requires covers to be removed so that terminals or connections that are live, or can be made live, are exposed, precautions should be taken to prevent unauthorised approach to or interference with live parts. This may be achieved by keeping the work area under the immediate surveillance of an employee or by erecting a suitable barrier, with Approved Notices displayed warning against approach and interference. When live terminals or site barriers are being adjusted, only approved insulated tools should be used. Additional precautions may be required because of the nature of any hazardous process or special circumstances present at the site of the work. Work on live equipment should only be undertaken where it is unreasonable in all the circumstances for it to be made dead.

E3.5 Operation of switchgear

The operation of switchgear should only be carried out by a Competent Person after he/she has obtained full knowledge and details of the installation and the effects of the intended switching operations. Under no circumstances must equipment be made operable by hand signals or by a prearranged time interval.

▼ **Figure E3.1**   Model form of permit-to-work (front)

PERMIT-TO-WORK

1. ISSUE   No............................

To ....

The following apparatus has been made safe in accordance with the Safety Rules for the work detailed on this Permit to-Work to proceed:

....

....

....

TREAT ALL OTHER APPARATUS AS LIVE
Circuit Main Earths are applied at

....

....

....

Other precautions and information and any local instructions applicable to the work, notes 1 and 2.

....

....

....

The following work is to be carried out: ....

....

....

Name (Block capitals) ....

Signature ....

Time ....   Date ....

▼ **Figure E3.1** *continued* (back)

2. RECEIPT
(Note 2)

I accept responsibility for carrying out the work on the Apparatus detailed on this Permit-to-Work and no attempt will be made by me, or by the persons under my charge, to work on any other Apparatus.

Name (Block capitals) ....

Signature ....

Time .... Date ....

3. CLEARANCE
(Note 3)

All persons under my charge have been withdrawn and warned that it is no longer safe to work on the Apparatus detailed on this Permit-to-Work, and all Additional Earths have been removed.

The work is complete*/incomplete*

All gear and tools have*/have not* been removed

Name (Block capitals) ....

Signature ....

Time... Date ....

....

....

....

*Delete words not applicable

4. CANCELLATION
(Note 3)

This Permit-to-Work is cancelled.

Name (Block capitals) ....

Signature ....

Time .... Date ....

▼ **Figure E3.1** *continued*

5. DIAGRAM

The diagram should show:

(a) the safe zone where work is to be carried out
(b) the points of isolation
(c) the places where earths have been applied, and
(d) the locations where 'danger' notices have been posted.

Notes on Model Form of Permit-to-Work

1. ACCESS TO AND WORK IN FIRE PROTECTED AREAS

Automatic control

Unless alternative Approved procedures apply because of special circumstances then before access to, or work or other activities are carried out in, any enclosure protected by automatic fire extinguishing equipment:

(a) The automatic control shall be rendered inoperative and the equipment left on hand control. A Caution Notice shall be attached.

(b) Precautions taken to render the automatic control inoperative and the conditions under which it may be restored shall be noted on any Safety Document or written instruction issued for access, work or other activity in the protected enclosure.

(c) The automatic control shall be restored immediately after the persons engaged on the work or other activity have withdrawn from the protected enclosure.

2. PROCEDURE FOR ISSUE AND RECEIPT

(a) A Permit-to-Work shall be explained and issued to the person in direct charge of the work, who after reading its contents to the person issuing it, and confirming that he understands it and is conversant with the nature and extent of the work to be done, shall sign its receipt and its duplicate.

(b) The recipient of a Permit-to-Work shall be a Competent Person who shall retain the Permit-to-Work in his possession at all times whilst work is being carried out.

(c) Where more than one Working Party is involved a Permit-to-Work shall be issued to the Competent Person in direct charge of each Working Party and these shall, where necessary, be cross-referenced one with another.

3. PROCEDURE FOR CLEARANCE AND CANCELLATION

(a) A Permit-to-Work shall be cleared and cancelled:

(i) when work on the Apparatus or Conductor for which it was issued has been completed;

(ii) when it is necessary to change the person in charge of the work detailed on the Permit-to-Work;

(iii) at the discretion of the Responsible Person when it is necessary to interrupt or suspend the work detailed on the Permit-to-Work.

(b) The recipient shall sign the clearance and return to the Responsible Person who shall cancel it. In all cases the recipient shall indicate in the clearance section whether the work is 'complete' or 'incomplete' and that all gear and tools 'have' or 'have not' been removed.

(c) Where more than one Permit-to-Work has been issued for work on Apparatus or Conductors associated with the same Circuit Main Earths, the Controlling Engineer shall ensure that all such Permits-to-Work have been cancelled before the Circuit Main Earths are removed.

4. PROCEDURE FOR TEMPORARY WITHDRAWAL OR SUSPENSION

Where there is a requirement for a Permit-to-Work to be temporarily withdrawn or suspended this shall be in accordance with an Approved procedure.

Index

Electrical **excellence**

Expert publications

The IET is co-publisher of BS 7671 (IET Wiring Regulations), the national standard to which all electrical installations should conform. The IET also publishes a range of expert guidance supporting the Wiring Regulations.

You can view our entire range of titles including...

- BS 7671
- Guides
- Guidance Notes series
- Inspection, Testing and Maintenance titles
- City & Guilds textbooks and exam guides

...and more at:

www.**theiet**.org/electrical

ELECTRICAL STANDARDS ✚

Constantly up-to-date digital subscriptions

Our expert content is also available through a digital subscription to the IET's Electrical Standards Plus platform. A subscription always provides the newest content, giving peace of mind that you are always working to the latest guidance.

It also lets you spread the cost of updating all your books once new versions are released.

Going digital gives you greater flexibility when working with the Wiring Regulations and guidance, with an intuitive search function and instant linking between Regulations and guidance. You can also access the content on your desktop, laptop, mobile or tablet, making it easy to take the content out on site or read on the move.

Find out more about our subscription packages and choose one to suit you at:

www.**theiet**.org/esplus

IET Standards

Industry-leading standards

IET Standards works with industry-leading bodies and experts to publish a range of codes of practice and guidance materials for professional engineers, using its expertise to achieve consensus on best practice in both emerging and established technology fields.

See the full range of IET Standards titles at:

www.theiet.org/standards

The Institution of Engineering and Technology

IET Centres of Excellence

The IET recognises training providers who consistently achieve high standards of training delivery for electrical installers and contractors on a range of courses at craft and technician levels.

Using an IET Centre of Excellence to meet your training needs provides you with:

- Courses that have a rigorous external QA process to ensure the best quality training
- Courses that underpin the expertise required of the IET Electrical Regulations publications
- Training by competent and professional trainers approved by industry experts at the IET

See the current list of IET Centres of Excellence in your area at:

www.theiet.org/excellence